A TAXONOMIC REVISION OF THE GENUS ORIGANUM
(Labiatae)

LEIDEN BOTANICAL SERIES

Volume 4

Series ISBN 90-6021-462-5

This book can be cited as:

J.H. IETSWAART. 1980. A taxonomic revision of the genus Origanum (Labiatae). Leiden University Press (Leiden Botanical Series, vol. 4). XI + 153 pp., 36 figs., 6 tables.

In the Leiden Botanical Series will be published papers of a monographic nature from the entire field of botany (including its history, bibliography, and biography) which by their length (100 printed pages or more) are unsuitable for publication in journals. Information can be obtained from the editors, Rijksherbarium, Schelpenkade 6, Leiden, The Netherlands.

A TAXONOMIC REVISION OF THE GENUS ORIGANUM

(Labiatae)

J.H. IETSWAART

Vrije Universiteit, Amsterdam

1980

LEIDEN UNIVERSITY PRESS

THE HAGUE / BOSTON / LONDON

Distributors:

for the United States and Canada

Kluwer Boston, Inc.
190 Old Derby Street
Hingham, MA 02043
USA

for all other countries

Kluwer Academic Publishers Group
Distribution Center
P.O.Box 322
3300 AH Dordrecht
The Netherlands

Published with financial support from the Netherlands Organization for the Advancement of Pure Research (Z.W.O.).

ISBN-13: 978-90-6021-463-3 e-ISBN-13: 978-94-009-9156-9
DOI: 10.1007/ 978-94-009-9156-9

Aan mijn ouders
Voor Else

CONTENTS

SUMMARY

The present study deals with the systematics and taxonomy of the genus *Origanum* (*Labiatae*, *Saturejeae*). As this difficult genus was never before monographed, a revisional study was much needed. The data presented are mainly based on the study of herbarium specimens and in some cases of living ones. The picture was completed, as far as possible, with data from various literature sources.

A short survey is given of the taxonomic history of *Origanum*, which goes back as far as Linnaeus, and shows that genus and species concepts of various authors have much differed.

A morphological outline of *Origanum* is given, from which it can be concluded that most generic characters are rather variable. *Origanum* is characterized in the following ways. Medium sized, subshrubby *Labiatae*, rich in volatile oils, with subsessile, ovate, glandular punctate leaves and paniculate inflorescences; few flowered verticillasters arranged in (dense) spikes with distinct, often coloured, bracts; calyces variable: 5-toothed, subregular or 2-lipped or 1-lipped, with developed or reduced teeth; corollas 2-lipped, sometimes saccate or flattened.

Origanum is compared with related genera found in the subtribes *Melissinae* and *Thyminae* within the tribe *Saturejeae*. One conclusion is that there are no arguments to maintain these subtribes. Further it can be concluded that *Origanum*'s nearest relatives are *Thymus*, *Satureja* and *Micromeria*. In the sections *Campanulaticalyx* and *Elongatispica*, *Origanum* comes near to the latter genus. The genera *Satureja* and *Micromeria*, which together contain the bulk of the species in the group, are in need of a revision. When this is carried out it may become clear that several genera should be redefined, including possibly *Origanum*.

The genus is divided into 10 sections, of which two are new and one transferred from another genus. In all 38 species are recognized. Specific differences are found in the indumentum and in the size and/or shape of spikes, bracts, calyces, corollas, and filaments. These and other characters are uniformly included in the descriptions given. In two species infraspecific taxa are listed. In addition 17 hybrids are recognized, of which four are new and three others were previously described as species. For six taxa a new status is introduced (in one case in a new combination), while two new combinations are made, one species name is validated, and one new name is given. Type specimens are recorded for all taxa and identification keys to all taxa are given. Important characters are picutred for all species and infraspecific taxa. Distribution maps are given.

The chromosome number of four species of *Origanum* is known at the moment. In all cases (apart from a few counts for *O. vulgare*) the number 2n = 30 is established.

Gynodioecy occurs in the species of five sections.

Most *Origanum* species (c. 70 %) are found in the East Mediterranean subregion, while a few species occur in the West Mediterranean subregion. Most species occupy (rather) small areas: c. 70 % is endemic to one island or mountain (group). Only *O. vulgare* has a very large area, ranging from the Azores to Taiwan. *Origanum* species usually inhabit mountain regions and rocky places with calcareous stone.

Though hybridization is frequently found in *Origanum*, hybrids do not usually occur in large numbers. It is postulated that not only intra-, but also inter-generic hybrids occur in this group of *Saturejeae*.

In a hypothesis for speciation hybridization is seen as the most important way of origin of *Origanum* species. This hybridization can have taken place between species of *Origanum* as well as between *Origanum* species and species from related genera.

Origanum species are generally rich in volatile oils containing considerable quantities of carvacrol and thymol.

Since ancient times species of *Origanum* are used as medicinal herbs but nowadays this is of minor importance. In the course of time the use of *Origanum* species as culinary herbs has become more important. In recent times several species of *Origanum* have also been used as ornamentals. Two *Puccinia* species and a strain of alfalfa mosaic virus are found as parasites on *Origanum*.

ACKNOWLEDGEMENTS

It is a pleasure to have the opportunity here to express my gratitude to all persons and institutes which were helpful with this work in one way or another.

Prof. Dr. M. Vroman gave me the opportunity to carry out this study at the Department of Biology at the Vrije Universiteit. He supported this study with interest and also contributed much to the final version of this revision. Prof. Dr. C. Kalkman (Rijksherbarium, Leiden) shared generously his and the Leiden experience in writing monographs, resulting in much constructive criticism and invaluable advice.

In the first stage of the work some useful advice was obtained from Prof. Dr. J. Lanjouw (formerly State University, Utrecht) and Prof. Dr. V.H. Heywood (University of Reading).

Drs. N.J. van Strien produced several fine illustrations, later on Mr. G.W.H. van den Berg continued this work. Mr. J.H. Huysing made many indispensable photographs. Other technical assistance was received from amongst others, Miss A.E.G. van Beek and Miss M. Duisterhof.

Miss C.T.M.J. Keulemans was of great help with search for, often obscure, literature.

Former undergraduate students, B.N. Blom, A. Fokkinga and J. Herder, each contributed a part to the overall picture of Origanum as presented here.

A number of plants were cultivated in the botanical garden of the Vrije Universiteit under supervision of Mr. D. Smit (curator) and Mr. P. Donk (assistant curator).

Prof. Dr. C. Wilkinson kindly corrected the English text. Drs. D.C. Ietswaart and Dr. J.F. Veldkamp took care of the Latin in some diagnoses.

The difficult type-work was done by Mrs. P.F.M. Ohr.

The following persons are gratefully mentioned here for various assistance: Prof. Dr. L. Boulos (formerly University of Tripoli, now National Research Institute, Cairo), Dr. A. Danin (University of Jerusalem), Dr. P.H. Davis (University of Edinburgh), Dr. S. Ecoyomides (Agricultural Research Institute, Cyprus), Dr. I. Fertig-Grünberg (University of Jerusalem), Prof. Dr. W. Greuter (formerly University of Genève, now Botanisches Museum, Berlin), Dr. J.C. Hedge (University of Edinburgh), Dr. H. Maarse (Centraal Instituut voor Voedingsonderzoek, Zeist), Dr. K. Meyer (University of Jena), the late Dr. P. Mouterde (University of Beirouth), Prof. Dr. K.H. Rechinger (formerly Naturhistorisches Museum, Wien), Dr. W.T. Stearn (formerly British Museum, London).

Last but not least many thanks should be given to the directors and curators, and private owners of the herbaria mentioned below. Their kind permission for and assistance with sending herbarium specimens on loan (and/or photographs of type specimens) were very essential to this study. From the following institutes and private persons herbarium specimens were received:
Botanisches Museum, Berlin; British Museum (Natural History), London; Museum of Natural History (Department of Botany), Budapest; Dr. Buttler, München; Department of Botany (Faculty of Sciences), Cairo; Istituto di Botanica, Orto Botanico, Catania; Botanical Institute of The University, Coimbra; Royal Botanic Garden, Edinburgh; Herbarium Universitatus Florentinae (Istituto Botanico), Firenze; Conservatoire et Jardin botaniques, Genève; Dr. Huber-Morath, Basel; Institut für Specielle Botanik und Herbarium Haussknecht, Jena; The Herbarium and Library, Kew; Goulandris Natural History Museum, Kifissia; Rijksherbarium, Leiden; The Linnean Society, London; Instituto "Antonio José Cavanilles", Madrid; Fielding Herbarium, Druce Herbarium (Department of Botany) Oxford; Muséum National d'Histoire Naturelle, Laboratoire de Phanérogamie, Paris; Erbario dell'Istituto Botanico, Palermo; Academy of Natural Sciences, Philadelphia; Universitatis Carolinae, Facultatis Biologicae Scientiae Cathedra, Praha; Dr. Sorger, Wien; Naturhistoriska Riksmuseum (Botanical Department), Stockholm; Instituut voor Systematische Plantkunde, Utrecht; Naturhistorisches Museum, Wien; Botanisches Institut und Botanischer Garten der Universität, Wien.

I. GENERAL CHAPTER

I.1. INTRODUCTION

In chapter I a general picture is given of the genus *Origanum*. This means that various data are brought together: abstracts from literature as well as my own observations and hypotheses. The picture is not complete; in the field of chemistry (volatile oils, flavonoids etc.) and cytology (chromosomes) much work has still to be done.

I.2. TAXONOMIC HISTORY OF ORIGANUM

In this short survey of the taxonomic history of the genus *Origanum* only the most salient points will be mentioned. The genus was described by Linnaeus (1754), referring to Tournefort, in the 5th edition of Genera Plantarum. This work, combined with his Species Plantarum (1753), gives a clear picture of Linnaeus' concept of the genus, which can be summarized as follows: *Labiatae* with flowers in more or less dense spikes, bracts conspicuous and often coloured, calyces with five equal teeth or 2-lipped or lower lip reduced, corollas 2-lipped. The number of species described by Linnaeus is given in table 1, together with those of several other authors. Miller (1754) described three species under the genus *Majorana*. Later on (1768), however, he abandoned this concept and placed the three species under *Origanum*. Another species of *Origanum* was transferred by Gleditsch (1764) to the genus *Amaracus*. Vogel (1840, 1841) recognized *Majorana* and *Amaracus* as subgenera within the genus *Origanum*. Rafinesque (1836) and Scheele (1843) introduced some other generic names for parts of the Linnaean genus *Origanum*, which never gained acceptance. Kuntze (1867, 1891) united the genera *Origanum* and *Thymus*, first under *Thymus*, later under *Origanum*. None of these views have been followed. In his earlier days Bentham (1834) recognized three genera: *Amaracus*, *Majorana* and *Origanum*. Later on (1848) and in Bentham & Hooker (1876), he switched to the Linnaean genus concept. In 1848 he described four sections: *Amaracus*, *Majorana*, *Origanum* and a completely new one, *Anatolicon*. Briquet (1895) accepted three separate genera in which he adopted some of Bentham's sections and also described some new ones: *Schizocalyx*, *Holocalyx* and *Chilocalyx* (all in the genus *Majorana*). Up to the most recent times both the three genera and the one genus concept have been in use. Among the authors who followed the first concept are: Moench (1794), Bornmüller (1917), Hayek (1931), Rechinger (e.g. 1943a), Wolf (1954), Wunderlich (1967) and Zohary (1973). In the Paris Code 1956 both *Amaracus* and *Majorana* have been accepted as nomina generica conservanda with *A. tomentosus* and *M. hortensis* respectively as type species. Among additional authors who recognized the genus *Origanum* in a broad sense are: Lamarck (1797), Sibthorp & Smith (1826), Willkomm & Lange (1868), Boissier (1879), Nyman (1881, 1890), Halácsy (1902), Post & Dinsmore (1933), Mouterde (1935), Davis (1949), Thiebaut (1953), El-Gazzar & Watson (1970), Fernandes & Heywood (1972).

The most complete survey known up to now of the East-Mediterranean species was given by Boissier in his "Flora Orientalis" (1879). An up to date treatment of the genus for this area will appear in the "Flora of Turkey" (Ietswaart, 1981). The European species of the genus have been dealt with by Nyman (1881) in the "Conspectus Florae Europaeae" and by Tutin et al. (1972) in the "Flora Europaea".

A revision and a survey of the genus as a whole have never been given before. Some of the new sections of which it was intended to present them first in this revision, have been published recently elsewhere (Ietswaart, 1980).

Table 1. Most important authors who described alone or with a co-author at least 2 species or hybrids in the genera *Amaracus*, *Majorana* and/or *Origanum*. The year(s) of publication is (are) given with each author, and also the number of those taxa recognized in the present monograph as species or hybrids.

author + year(s) of publication	number of species and hybrids described	number of those recognized
LINNAEUS, 1753, 1763	11	5
MILLER, 1768	4	1
SALISBURY, 1796	4	–
MOENCH, 1794, 1802	2	–
WILLDENOW, 1800	2	–
STOKES, 1812	4	–
LINK, 1809, 1822	4	–
BENTHAM, 1834, 1836	3	3
RAFINESQUE, 1836, 1840	2	–
SAVI, 1840	2	–
VOGEL, 1840, 1841	5	1
KOCH, 1848	4	–
BOISSIER, 1844, 1846, 1854, 1859, 1879	11	8
POST, 1893, 1895	2	2
BORNMÜLLER, 1898, 1917	2	2
DAVIS, 1949, 1951, 1956	8	8
RECHINGER, 1938, 1943b, 1952, 1954, 1961	5	2
DANIN, 1967, 1969	2	2
MOUTERDE, 1935, 1973	4	4

I.3. CONCEPT OF ORIGANUM IN THIS REVISION

In the present revision the original broad *Origanum* concept of Linnaeus has been accepted. In my opinion this is preferable to the most commonly accepted alternative, a division into three genera, as worked out by Briquet. Five arguments are presented in favour of this sensu lato concept. First of all, there are several characters that the species of *Origanum* s.l. have in common (see I.4). Secondly, several characters mentioned by Briquet as being exclusive for one of his three genera do not hold absolutely (see also I.4). A third argument is found in the occurrence of c. 13 hybrids between species and subspecies from different sections of *Origanum* s.l. These hybrids combine characters of two different genera sensu Briquet. This in the fourth place, can also be said of some groups of species (distinguished in this revision as sections). The origin of these species most probably must be found in processes of hybridization (see I.11). A last argument is derived from the study of chromosomes. Only a few *Origanum* species have been investigated until now, but at least one or two from each of the three genera of Briquet. In all cases (except a few populations of *O. vulgare*) the number 2n = 30 has been established (I.8).

A diagnosis of *Origanum* with respect to related genera will be given in subchapter I.5. In the present revision of the genus 10 sections are recognized: *Amaracus*, *Anatolicon*, *Brevifilamentum*, *Longitubus*, *Chilocalyx*, *Majorana*, *Campanulaticalyx*, *Elongatispica*, *Origanum* and *Prolaticorolla*. In I.6 some criteria are given concerning the delimitation of sections, species, subspecies and varieties. Recognized are 38 species, one of which with 6 subspecies and another with 3 varieties. In addition 17 taxa of hybrid origin are distinguished.

I.4. MORPHOLOGICAL CHARACTERS OF ORIGANUM

Underground parts

The subterranean parts of all *Origanum* species are more or less woody. Many species (especially those from the East Mediterranean area) have very thick, woody roots.

Stems

In nearly all Mediterranean species the lower portions of the epigeal parts are also woody and persistent (subshrubs). In nearly all species stems are erect, or ascending. *O. vulgare* is the only species which forms long superficial runners. The non-woody and erect parts of the stems often measure only 1 – 2 mm in diameter but can reach 5 mm in diameter. Usually there are many stems (up to several tens) in one plant. The lengths of the stems are rather variable. Some species have stems between 10 and 30 cm long. In the majority of the species the stem length is 30 – 60 cm. Several species however, can have stems from 60 cm to more than 1 metre. In herbarium specimens these last lengths are of course seldom found. In cross-section the stems are roundish or slightly quadrate. The internodes show a mean length of c. 2.5 cm, but vary from c. 1 – 4 cm. The stems usually bear side branches. These branches are usually found in the upper 1/4 – 1/2, seldom as low as 4/5. In some species of the sections *Amaracus*, *Anatolicon*, *Brevifilamentum* and *Longitubus* branches are absent. In the other species of these groups the stems usually bear a few pairs of unbranched laterals. In the section *Campanulaticalyx* usually many pairs of short primary branches are present. In all other sections the species are characterized by rather long branches of the first order, branched again up to the third order. The mean value of the number of pairs of branches of the first order is c. 6, while it can be as much as 30. The length of the primary branches, measured without the spikes, varies from 0.1 cm (subsessile spikes) to 35 cm. Usually the erect part of the stems dies down (almost) to the base after each growing season. In some Eu-Mediterranean species erect stem parts remain for two or more seasons on the plants. Most stems are hairy, at least at the base. In only a few species the stems are glabrous, in which case the whole plant usually is. When the stems are only slightly hairy at the base the rest of the plant is usually (almost) glabrous. When the stems are completely hairy, hairs are usually found in all other parts of the plant. In all species hairs are simple, except in *O. dictamnus* were they are branched. The length of the hairs varies from less than 0.1 to 3 mm. Within a species hairs can vary considerably in length and number. It is notable that in some species with glabrous stems and leaves the young sprouts are thickly covered with hairs (e.g. *O. sipyleum*).

Leaves

Leaves are (sub)sessile in several species, in many others petiolate, especially in the lower nodes. Petioles on the lower parts of the stems are usually c. 1/4 as long as the blades, but they can reach half this length. Of course the leaf blades also vary much in shape and size. In some species they are nearly as broad as long, but in other species they can be 2 (– 2.5) times longer than wide. The mean value for the leaf length/width ratio is 1.3. The leaves vary in length from 2 – 40 mm and in width from 2 – 30 mm. The smallest leaves are found in a few species (e.g. *O. microphyllum*) on short secondary branches. The indumentum on the leaves does not differ from that on the stems, but often the hairs on the leaves are somewhat shorter. On the leaves two types of glands occur, sessile and stalked ones. These also occur on stems, bracts, calyces and corollas. Little work has been done on these glands on the different species. Only the number of the sessile glands on the leaves has been counted. A considerable variation of 50 – 2500 per square cm, can be summarized as a mean value of 650 per square cm.

When the leaves are more or less glabrous they nearly always are glaucous (i.e. covered with a thin waxy layer). In this case the leaves are usually also more or less leathery, which is caused by a thickened cuticula. This, just as hairs, waxy layers and glands, must be seen as a way to diminish evaporation in *Origanum*. In a very few species revolute leaf margins are found. The adaptation of linear leaves, as found in many species of related genera, does not occur in *Origanum*.

Inflorescences

In principle each stem and branch bears a spike. This means that depending on the number of branches more or less paniculate inflorescences (panicles) occur. Only in one case *(O. onites)* do all spikes lie in one level and form a false corymb. The spikes vary in size from 2 × 2 to 45 × 25 mm (measured without the flowers). The smallest spikes are found in the sections *Chilocalyx* and *Campanulaticalyx*, and contain no more than 2 – 4 very small bracts. The largest spikes occur in the sections *Amaracus*, *Brevifilamentum* and *Longitubus*. In these sections and in the section *Anatolicon* the spikes are usually nodding. Usually the spikes are compact and the bracts imbricate. In the section *Majorana* the spikes are even very dense. In the sections *Elongatispica* and *Campanulaticalyx*, however, the bracts are only slightly or not imbricate, and the spikes loose. The mean number of pairs of bracts in a spike is 7, while this number can vary from 2 to 40. In shape the bracts are usually roundish or (ob)ovate, more seldom lanceolate. The larger bracts are usually somewhat hemispherical or boat-shaped. In the section *Majorana* the bracts enclose the calyces marginally. The other small bracts are (nearly) flat. In some cases (sections *Chilocalyx* and *Campanulaticalyx*) the leaves gradually change into bracts. The bracts vary greatly in relative and absolute size. They can be nearly as long as the calyces (2 – 4 mm), and up to 3 times longer than the calyces and c. 25 mm long. The mean length/width ratio for the bracts is 1.7 and shows a variation from 0.9 to 4. Small bracts are usually leaf-like in texture, colour and indumentum, while larger bracts are thinner and somewhat papery (membranous). These larger bracts are (partly) purple coloured or yellowish green, and usually glabrous or slightly hairy. This type is found amongst others in the sections *Amaracus* and *Anatolicon*, but also in one species of the section *Prolaticorolla* and in 2 subspecies of *O. vulgare* (section *Origanum*). Most *Origanum* species have only 2, sessile flowers in a verticillaster, but in the sections *Brevifilamentum* and *Longitubus* usually more flowers, up to 16, are found. Here c. 1 mm long penduncles occur.

Calyces

The calyces are amongst the most variable parts within the genus *Origanum*. Yet they have one character in common: in outline they are straight. In the sections *Elongatispica*, *Origanum* and *Prolaticorolla* the calyces are regularly 5-toothed for c. 1/3 of their total length, tubular, and 2.5 – 6 mm long. In some species with a calyx of this type the calyx teeth close in the fruiting stage. In the section *Campanulaticalyx* the calyces are sub-regularly 5-toothed for c. 2/5, more or less campanulate, and 1.5 – 7.5 mm long. In the section *Majorana* the calyces are 1-lipped for c. 9/10, flattened, and 2 – 3.5 mm long. This type of calyx is usually (angulate) obovate. In the section *Chilocalyx* the calyces are usually 2-lipped (seldom 1-lipped) for c. 1/5, (nearly) tubular and 1.5 – 3 mm long. The teeth in upper and lower lips are very small (seldom absent). In the section *Amaracus* the calyces are 1- or 2-lipped for c. 2/5 – 3/5, tubular, and 4 – 8 mm long. When teeth are present they are usually small. In the remaining sections, *Anatolicon*, *Brevifilamentum* and *Longitubus* the calyces are 2-lipped for 2/5 (seldom 1/5 or 1/2), tubular, and 4 – 12 mm long. The teeth in upper and lower lips are usually well developed. Teeth in all types of calyces are deltoid or triangular, rarely acuminate. In the species with 2-lipped calyces the shape of the calyx teeth can vary to some extent. In all types of calyces the throats are pilose, except the 1-lipped ones, in which the throats are (slightly or) not pilose.

Corollas

The most common type of corolla found in the genus has a straight tube, is 2-lipped for c. 1/3 (or 1/4), is 3 – 14 mm long, and has the lips at wide angles to the tube. This type is found in the sections *Anatolicon*, *Origanum*, and *Elongatispica*. In the sections *Prolaticorolla* and *Brevifilamentum* a similar type is found, which is 2-lipped for 1/6 respectively 1/5, while the length varies from 7 to 16 mm. In the latter section the tube is often slightly curved. This also can be said of *O. amanum* (the only species in the section *Longitubus*), where the corolla is 2-lipped for 1/7, while the lips stand at nearly right angles to the tube. This corolla is the longest in the genus (up to 40 mm). The corollas in the section *Amaracus* differ from the common type in having a sac-like protuberance about halfway at the underside. The lengths are here 8 to 17 mm. In the section *Chilocalyx* the corollas are 2-lipped for c. 2/5 and 2 – 7 mm long. This type is also found in two species of the section *Campanulaticalyx*. The corolla of the third species in this section is of the common type. In the section *Majorana* corollas occur of the same type as in the section *Chilocalyx*, but they are flattened. The colour of the corollas is white, pink or purple.

Stamens and ovaries

The staminal filaments can be subequal in length (e.g. section *Anatolicon*) or very unequal (e.g. section *Brevifilamentum*). They can protrude far beyond the lips of the corolla (e.g. section *Amaracus*), protrude slightly (e.g. section *Chilocalyx*), or be included (section *Longitubus*). Further the filaments can stick straight out (e.g. section *Origanum*), ascend under the upper lip of the corolla (e.g. section *Amaracus*), or diverge (e.g. section *Majorana*). In length the staminal filaments vary much: from 0.5 to 14 mm. The filaments are usually glabrous, except in two species of the section *Campanulaticalyx*, where they are pilosellous. The styles usually protrude to the same extent as the stamens do, and can be up to 22 mm long (seldom up to 40 mm). The two stigma lobes are usually slightly unequal in size and c. 0.5 mm long. The large variation in the shape and/or size of thc styles, stamens and corollas, as well as the upside-down position of the latter in several species, in principle must be considered as adaptations to various pollinating insects.

Fruits

The nutlets, which have been seen from some species only, are not much different. They are ovoid, brown, 1 – 1.5 mm long and c. 0.5 mm wide, and slightly acute to apiculate at the base.

Pollen

Pollen grains of species in the sections *Amaracus*, *Anatolicon*, *Brevifilamentum* and *Longitubus* have been studied. All were hexacolpate and suboblate or oblate-spheroidal, varying in length from 30 to 50 µm and in width from 33 to 55 µm. These data agree with those of Waterman (1960) and Wunderlich (1967).

Summarizing, *Origanum* can be characterized as follows. Subshrubby *Labiatae* with several erect, medium sized stems per specimen, and with glandular punctate, ± ovate leaves. Few (sub)sessile flowers in a verticillaster. Verticillasters arranged in (dense) spikes with differentiated (coloured) bracts. Inflorescences more or less paniculate. Calyces straight, very variable: regularly 5-toothed, 2- or 1-lipped for 9/10 to 1/5; tubular, campanulate or flattened; throats pilose or not. Corollas variable: 2-lipped for 2/5 to 1/7, gibbous (saccate) or not, sometimes flattened; tube straight or slightly curved. Stamens subequal to very unequal in length; ascending, straight or divergent, extensively protruding to included.

I.5. COMPARISON OF ORIGANUM WITH RELATED GENERA

In table 2 four important classifications of the *Labiatae* are given. These will not be compared and discussed here as a whole, because this would be far beyond the scope of this study. In all systems but one the place of *Origanum* is the same, in one of the subtribes of the tribe *Saturejeae*. The partly dissenting concept of Briquet concerning *Origanum* will be discussed below. The system of Briquet is the most elaborated and most complete one, so it seems the best starting point for a discussion about the delimitation of *Origanum* with respect to other genera. Before coming to this it must be stated that Briquet's distinction of the subtribes *Melissinae* and *Thyminae* is of no value. The most important difference between these subtribes is the ascent of the stamens under the corolla upper lips in the *Melissinae* against the stamens sticking straight out in the *Thyminae*. Both types, with some others, are found in the genus *Origanum*. Further it can be stated that *Origanum*'s nearest relatives are found in the *Thyminae* as well as in the *Melissinae*. Also Wunderlich could not give any real difference between the two subtribes from the field of pollen morphology, development of the seed, and morphology of the full grown seed. Henceforth *Origanum* will be compared with the genera of both subtribes without further distinction. These genera are: *Acinos*, *Bystropogon*, *Calamintha*, *Ceranthera*, *Clinopodium*, *Conradina*, *Cunila*, *Hedeoma*, *Hedeomoides*, *Kurzamra*, *Melissa*, *Micromeria*, *Monardella*, *Pogogyne*, *Pycnanthemum*, *Saccocalyx*, *Satureja* s.s., *Thymbra*, *Thymus* (incl. *Coridothymus*), *Zataria* and *Ziziphora*. All genera are taken here sensu Briquet except *Satureja*, which, in accordance with the Flora Europaea treatment, is split into *Acinos*, *Calamintha*, *Clinopodium*, *Micromeria* and *Satureja* s.s.

Table 2. Four different classification systems of the *Labiatae*. Under I etc. are given the subfamilies, under 1 etc. the tribes, and under A etc. the sub-tribes. The lowest taxon (taxa) under which *Origanum* is placed is (are) underlined. The system of Wunderlich has somewhat been adapted to make it more easily comparable with the other 3 systems, but by no means with the intention of introducing new taxa.

BENTHAM & HOOKER 1876	BRIQUET 1895	MELCHIOR 1964	WUNDERLICH 1967
1. *Ocimoideae*	I. *Ajugoideae*	I. *Prostantheroideae*	I. *Prostantheroideae*
A. *Euocimeae*	1. *Ajugeae*	II. *Ajugoideae*	II. *Ajugoideae*
B. *Lavanduleae*	2. *Rosmarineae*	III. *Rosmarinoideae*	III. *Scutellarioideae*
2. *Satureineae*	II. *Prostantheroideae*	IV. *Ocimoideae*	IV. *Stachyoideae*
A. *Pogostemoneae*	III. *Prasioideae*	1. *Ocimeae*	1. *Prasieae*
B. *Menthoideae*	IV. *Scutellarioideae*	A. *Hyptidinae*	2. *Marrubieae*
C. *Melisseae*	V. *Lavanduloideae*	B. *Ociminae*	3. *Hypogomphieae*
D. *Lepechinieae*	VI. *Stachyoideae*	C. *Plectranthinae*	4. *Stachydeae*
3. *Monardeae*	1. *Marrubieae*	V. *Catopherioideae*	A. *Melittinae*
4. *Nepeteae*	2. *Perilomieae*	VI. *Lavanduloideae*	B. *Lamiinae*
5. *Stachydeae*	3. *Nepeteae*	VII. *Prasioideae*	5. *Pogostemoneae*
A. *Scutellarieae*	4. *Stachydeae*	VIII. *Stachyoideae*	V. *Saturejoideae*
B. *Melitteae*	A. *Prunellinae*	1. *Pogostemoneae*	1. *Nepeteae*
C. *Marrubieae*	B. *Melittinae*	2. *Saturejeae*	2. *Prunelleae*
D. *Lamieae*	C. *Lamiinae*	A. *Perillinae*	3. *Glechoneae*
6. *Prasieae*	5. *Glechoneae*	B. *Menthinae*	4. *Saturejeae*
7. *Prostanthereae*	6. *Salvieae*	C. *Hyssopinae*	A. *Melissinae*
8. *Ajugoideae*	7. *Meriandreae*	D. *Thyminae*	B. *Hyssopinae*
	8. *Monardeae*	E. *Saturejinae*	C. *Thyminae*
	9. *Hormineae*	(*Melissinae*)	D. *Menthinae*
	10. *Lepechinieae*	3. *Lepechinieae*	E. *Perillinae*
	11. *Saturejeae*	4. *Hormineae*	5. *Rosmarineae*
	A. *Melissinae*	5. *Perilomieae*	6. *Lavanduleae*
	B. *Hyssopinae*	6. *Marrubieae*	7. *Hormineae*
	C. *Thyminae*	7. *Nepeteae*	8. *Monardeae*
	D. *Menthinae*	8. *Stachydeae*	9. *Salvieae*
	E. *Perillinae*	A. *Melittinae*	10. *Meriandreae*
	12. *Pogostemoneae*	B. *Lamiinae*	11. *Lepechinieae*
	VII. *Ocimoideae*	C. *Prunellinae*	12. *Elsholtzieae*
	A. *Hyptidinae*	9. *Glechoneae*	13. *Ocimeae*
	B. *Plectranthinae*	10. *Meriandreae*	A. *Hyptidinae*
	C. *Ociminae*	11. *Monardeae*	B. *Plectranthinae*
	VIII. *Catopherioideae*	12. *Salvieae*	C. *Moschosminae*
		IX. *Scutellarioideae*	VI. *Catopheroideae*

Before starting with the delimitation of *Origanum* with respect to the genera mentioned a few remarks will be made about the morphology of the group. Generally it can be stated that most genera show as large a variation in several characters as *Origanum* does. The *Labiatae* in question can be annual or perennial herbs or (sub)shrubs. Stems can be short and decumbent or relatively long and erect. The main stems can be branched or not. Leaves are usually small (1 – 2 cm long) and sessile or shortly petiolate, but some times relatively large (up to 10 cm and more long) and long-petiolate. The leaf ratio can be c. 1.5, but is often 5 and more. The leaf margin can be revolute or not. The flowers can occur in more or less pedunculate cymes, or in verticillasters which are shortly or not pedunculate. In both types 2 to many flowers can be present on a node. The verticillasters can be placed singly along the stems and branches, or can be arranged in terminal spikes. A combination of both types is often found. Spikes can be more or less globose or (long) ovoid, and also dense or loose. The inflorescences can be broadly paniculate through long branches or peduncles. The bracts can be similar to the leaves, or be very small and inconspicuous, or determine greatly the appearance of the spikes through size and colour. They can be ovate, or long and narrow, and densely imbricate or not. The calyces can be (sub)regularly 5-toothed or 2-lipped. The calyx teeth can be deltoid to triangular or very acuminate, and ciliate or glabrous. The throats can be pilose or not. In outline the calyces can be straight, curved, gibbous or inflated, sometimes they are dorso-ventrally flattened. The number of veins in the calyces is usually 10 – 13 (– 15), sometimes however 5 or 15 – 20. Corollas can be curved or gibbous. There are 2 or 4 stamens, while the thecae are rarely spurred or pilose at base.

The genus *Acinos* (c. 10 species in Europe, the Mediterranean, and C. Asia) is morphologically not very akin to *Origanum*. *Acinos* species have few flowers in shortly pedunculate cymes or in verticillasters. Bracts do practically not differ from the leaves. The calyces are 2-lipped, curved and gibbous.

The species in the genus *Bystropogon* (c. 10 species on the Canary Is., Madeira and in S. America) are characterized by multiflorous pedunculate cymes or by multiflorous verticillasters. The bracts are hardly or not distinct from the leaves. The calyces are always regularly 5-toothed. The genus is not closely related to *Origanum*.

This also can be said of the genus *Calamintha* (c. 7 species in W. Europe to C. Asia), with respect to its few-flowered verticillasters or shortly pedunculate cymes with undeveloped bracts.

Ceranthera (2 species in the southeastern part of N. America) is characterized by spurred thecae, and further by loose cymes, of which those at the top of the stems are united in spike-like inflorescences. The calyces are 2-lipped and the bracts somewhat different from the leaves.

In the species of *Clinopodium* (c. 10 in temperate Eurasia) dense, many-flowered verticillasters occur, while the calyces are 2-lipped and curved. Species in this genus can not easily be mistaken for *Origanum* species.

Conradina is another American genus (c. 4 species distributed like *Ceranthera*), with 2-lipped, acutely toothed calyces. The flowers are arranged in axillary verticillasters, and the leaves are linear.

Cunila (c. 15 species in N. America, Mexico, S. Brasil, Argentina and Peru) is characterized by regularly 5-toothed calyces and (small) flowers with 2 fertile stamens only. Often flowers occur in multiflorous pedunculate cymes, more seldom in verticillasters which are united into spikes. *C. origanoides* (L.) Briq. bears a superficial resemblance, in habit, with *Origanum* species.

Hedeoma (c. 30 species from Canada to Brasil) has 2 fertile stamens only. Its calyces are regularly 5-toothed or 2-lipped. The flowers are arranged in loose, few-florous verticillasters, of which the upper ones occur more or less in spikes.

Hedeomoides (c. 3 species in California) is much like *Pogogyne*, but it differs from the latter in having 2 fertile stamens.

Kurzamra is a monotypic genus, of which the only species, *K. pulchella* (Chile), has regularly 5-toothed calyces with thornlike teeth. Its verticillasters hold 2 flowers.

The species in the genus *Melissa* (c. 3 in Europe to C. Asia) are marked by 2-lipped calyces and curved corolla tubes. The verticillasters are few-flowered. As the other genera mentioned above, *Melissa* is quite different from *Origanum*.

Most *Origanum* species differ clearly from *Micromeria* species in 2- or few-flowered, not or hardly pedunculate verticillasters, which are all grouped closely together in (dense) spikes, and which are partly hidden by relatively large, purple or whitish green bracts. In the *Origanum* sections *Campanulaticalyx* and *Elongatispica* the differences are less obvious. Here the 2-flowered verticillasters are arranged in (very) loose spikes and the bracts are only slightly different from the leaves. In addition to the characters just mentioned differences must be sought in some non-floral characters: more or less paniculate inflorescences (section *Elongatispica*) and in the occurrence of ± roundish leaves (in both sections; against ± linear leaves in *Micromeria*). It must be stated that *Origanum* comes (very) close to *Micromeria* in sections *Campanulaticalyx* and *Elongatispica*. It is a common picture in the whole group of the *Melissinae*/*Thyminae* that a genus is linked to 1 or 2 other genera by "intermediate species". This should not be a reason for uniting all these genera, and as a consequence sometimes lines have to be drawn slightly artificially. The genus *Micromeria* contains the bulk of the species in the *Melissinae*/*Thyminae* (more than 100, occurring in most parts of the world). In fact *Micromeria* is the remnant group, when all other genera in the *Melissinae*/*Thyminae* have been delimitated.

In the genus *Monardella* (c. 20 species in western N. America) an obvious development of verticillasters in spikes with coloured bracts is found parallel with *Origanum*. In *Monardella*, however, the calyces are always equally 5-toothed, and the spikes more or less globose, not cylindrical. The leaf length/width ratio is often much more than 3.

Pogogyne (c. 3 species in southeastern N. America) is somewhat like *Monardella*, but all species are annuals. The verticillasters occur in spikes, with bracts, but often also isolated on the lower nodes.

Pycnanthemum (c. 17 species in N. America) differs from *Origanum* in (nearly) regularly 5-toothed calyces with glabrous throats, in many-flowered verticillasters arranged in globose spikes, and in the usually linear leaves.

Saccocalyx is a monotypic genus in N. Africa. *S. saturejoides* is easily characterized by its inflated calyces with c. 20 veins. In fact, these are the most important characters in which it differs from *Micromeria*.

What has been said above for *Origanum* and *Micromeria* also holds for *Satureja* s.s. This genus contains in its narrow circumscription c. 30 species occurring in many parts of the world. It is somewhat doubtful, whether *Micromeria* and *Satureja* can be sufficiently deliminated from one another. It should be mentioned here that a thorough revision of *Satureja*/*Micromeria* is much needed. When this is carried out possibly some new genera will have to be created and several old genera in the *Melissinae*/*Thyminae* have to be redefined. For *Origanum* this possibly could mean that the section *Campanulaticalyx* has to be excluded and placed in another genus. It does not seem rational to consider exclusion of the section *Elongatispica*.

The genus *Thymbra* (with 2 species in southeastern Europe to southwestern Asia) is mainly characterized by a dorso-ventrally flattened calyx, which is also found in the genus *Thymus* (subgenus *Coridothymus*) and in the genus *Origanum* (section *Majorana*). In the first two cases the calyces are 5-toothed and 2-lipped. In the section *Majorana* the calyces are 1-lipped, and no teeth are present. *Thymbra* further has few-flowered verticillasters arranged in dense spikes with relatively large, coloured bracts. The leaves are linear.

Thymus is another large genus in the group *Melissinae*/*Thyminae* with c. 50 species in temperate Eurasia. It also is a genus in which dense spikes with coloured bracts occur. *Thymus* differs from *Origanum* in a combination of characters, of which generally not all can be found when 2 species are compared. In *Thymus* solitary verticillasters occur as well as verticillasters arranged in spikes. In the latter case bracts are often not clearly differentiated. The verticillasters can be 2- to few-flowered as in *Origanum*. In *Thymus* the calyces are always 2-lipped, while the lower lip teeth are usually narrow, acuminate and ciliate (which is in *Origanum* seldom found). The staminal filaments are (sub)equal in *Thymus*, in *Origanum* they are often (very) unequal. *Thymus* species are often low plants with partly creeping stems, which are ramified at base, so that usually many equivalent stems are found and few or no branches. For this reason the typically paniculate *Origanum* inflorescence is seldom or never found in *Thymus*. Another non-floral difference between the 2 genera is found in the shape of the leaves. In *Thymus* the leaf length/width ratio is often more than 3, in *Origanum* usually less than 2. This is also valid for the *Thymus* species with large bracts and spikes. *Thymus* and *Origanum* probably converged in the same morphological direction from different parts of the same ancestral group (species like *Micromeria*). It is quite possible that species from the 2 genera have occasionally formed hybrids, which became species after some time. The same can be said about *Origanum* and *Micromeria* species.

Zataria is a monotypic genus not closely related to *Origanum*. *Z. multiflora* (Iran and Afghanistan), has a regularly 5-toothed calyx, which is 5-veined. The species

	ORIGANUM	ACINOS	BYSTROPOGON	CALAMINTHA	CERANTHERA	CLINOPODIUM	CONRADINA	CUNILA	HEDEOMA	HEDEOMOIDES	KURZAMRA	MELISSA	MICROMERIA	MONARDELLA	POGOGYNE	PYCNANTHEMUM	SACCOCALYX	SATUREJA	THYMBRA	THYMUS	ZATARIA	ZIZIPHORA
annuals		most								all			few	most	all			most				most
low plants	few	most		few			most		most	all	all		most	most	all			most		most	all	all
leaf ratio >3		few			all		all		most				most	all	all	most	all	most	all	most	most	most
leaf margins revolute	few						all						most	most				most		most		
flowers in pedunculate cymes			most	all	all		all	most	most				most					most		few	most	
flowers in verticillasters	all	all	most			all		most	most	all	all	all	most	all	all	all	all	most	all	most	most	all
many flowers at a node	few		all	few	all	all		most					most	most	most	all		most				
flowers in spikes only	all												few	all	few	all		few	all	most		most
spikes ± globose	few																			most		most
bracts prominently different from leaves	all												most	all	all	all		most	all	most	all	most
bracts large and purple	most													most					all	most		
calyces (sub)regularly 5-toothed	most		all					all	most		all		most	all		most	all	most			all	
calyces 2-lipped	most	all		all	all		all		most	all		all	most		all	few		few	all	all		all
calyx teeth (2 or 5) acuminate		most	most	most		most	all		most	all	all	all	most		all			most	all	all		
calyces curved or inflated		all				all							most				all	few				
calyces flattened	few																		all	few		
veins in calyces not 10-13(-15)																	all				all	
calyx throats pilose	most	all	all	all	all	all		all	most				most					most	most	all		all
corolla tube curved or gibbous	most							all				all										
stamens 2								all	all	all												all
thecae spurred or ciliate					all		all															

Table 3. Most important characters for delimitation of the genera in the *Melissinae*/*Thyminae*.

all = occurring in (nearly) all species; most = occurring in most species; few = occurring in a few species; (empty) = occurrance not found in literature nor in any species studied has more or less pedunculate cymes, while the bracts are slightly developed.

Ziziphora, the last genus to be discussed here, is one of the 4 genera in the group *Melissinae/Thyminae* in which there are only 2 stamens. The calyces are 2-lipped and the verticillasters more or less arranged in spikes. Bracts are more or less differentiated. All species are annuals.

In table 3 a summary is given of the most important characters which are used in the delimitation of the genera in the *Melissinae/Thyminae* and which have been

Table 4. Characters of the genus *Origanum*, and importance for delimitation of the genus, and the sections and species within it as discerned in the present revision:
+ + + : of most important weight, very frequently used
+ + : of much weight, often used
+ : of some weight, sometimes used
± : of slight weight, seldom used
− : of no weight, not used.

Characters	importance of character for delimination on the level of		
	genus	sections	species
1. any character of roots	−	−	−
2. length of stems	+	−	+
3. indumentum of stems and leaves	−	±	+ +
4. arrangements of branches	+ +	+	+
5. number of branches	+ +	+	+
6. length of branches	+ +	−	±
7. length of internodes	−	−	±
8. shape of leaves	+ +	−	+
9. length of petioles	+	−	−
10. number of sessile glands on leaves	+	−	+
11. arrangement of verticillasters	+ + +	±	−
12. shape of spikes	−	+ +	+
13. number of flowers in a verticillaster	±	+	−
14. length of peduncles	±	−	−
15. shape of bracts	+ +	+	+
16. size of bracts	+ +	+ + +	+ +
17. texture of bracts	+	+	−
18. colour of bracts	+	+	+
19. general shape of calyces	+ +	+ + +	+
20. shape of lips and teeth in calyces	+ +	+ +	+ + +
21. length of calyces	−	+	−
22. general shape of corollas	+ +	+ + +	+ +
23. shape of lips and lobes in corollas	+	+ +	+
24. colour of corollas	−	−	+
25. arrangement of stamens	−	+ +	±
26. length of staminal filaments	−	+ +	−
27. length of styles	−	+	−
28. any character of nutlets	−	−	−
29. any character of pollen	−	−	−
30. chromosome numbers	+	−	−
31. any chemical characters	−	−	−

partly discussed here. In table 4 the characters used for establishing the outer limits of *Origanum* are compared with those used for delimitation of the sections and the species within the genus.

I.6. CRITERIA FOR LIMITATION OF SECTIONS, SPECIES AND SUB-SPECIFIC TAXA

In this revision a section is understood to be a group of related species, which have more morphological characters in common with each other than with other species. Because of the large variation found in nearly all characters in the genus *Origanum*, no less than 10 sections had to be recognized. Of these, five had been described previously by various authors (see I.1), based on diverse characters. In this revision the sections are uniformly based on a number of characters, mainly concerning shape and/or size of spikes, bracts, calyces, calyx teeth, corollas and staminal filaments.

The 38 species in the genus are mainly discerned by the characters: shape and size of bracts, calyx teeth and corollas; the indumentum is also an important character (see table 4). All specimens belonging to one species always differ in the same combination of two or more characters from all specimens belonging to another species. When intermediates have been found linking two generally recognized species, they have been united into one species, except when too few specimens of the species and/or intermediates were available for study. In this case the current opinion has been adopted, and further study recommended.

Most infraspecific variation has not been named. A subspecies has been recognized only when all specimens from several local populations of a species were found to be different from the specimens in the "type population", but also specimens were found from several obviously intermediate populations. This has only been adopted for one species, of which many specimens were available for study. Also varieties have been named in one case only. Here specimens from different populations were distinct in at least one character, but not enough specimens were available to decide whether subspecies should be described or not.

I.7. GYNODIOECY

Gynodioecy is frequently found in species in the sections *Chilocalyx*, *Elongatispica*, *Majorana*, *Origanum* and *Prolaticorolla*. This became clear when studying herbarium specimens of the various species. Sometimes specimens were even found on which bisexual as well as female flowers occurred. Additionally some data relating to this subject will be given here from literature; all concern field observations of and breeding experiments with specimens of *O. vulgare* ssp. *vulgare*. Lewis & Crowe (1956) estimated that in populations of *O. vulgare* in western and northern Europe

30 – 50% of the plants have female flowers. They postulated that the gynodioecy system is controlled by two independent genes F and H. F causes abortion of the pollen and H is a dominant suppressor of F. The combinations with HH are selfincompatible, while the combination ffhh is probably lethal. They had to propose such a theory because the segregation patterns deviate from the Mendelian laws. Kheyr-Pour (1969) surveyed many populations in southern France, and found c. 4 – 18% of the plants to be female. He stated that it was rather difficult to determine these percentages exactly because *O. vulgare* plants make long running offsets, because specimens occur with male and female flowers together on one stem, and because "female plants" show bisexual flowers later on in the season. Valdeyron, Dommée & Valdeyron (1973) developed a computer simulation model for gynodioecy which fits the *O. vulgare* situation much better than other models. The model assumes the rare situation of a predominantly self pollinated population with a strong inbreeding depression. In this *O. vulgare* differs greatly, according to the authors, from, amongst others, *Thymus vulgaris*, in which gynodioecy is also found. The latter study has also been carried out with populations in southern France.

It can be concluded that the genetical basis of gynodioecy in *Origanum* seems to be rather complex. It would be very interesting to study other *Origanum* species in this respect.

I.8. CHROMOSOME NUMBERS

All *Origanum* taxa for which chromosome numbers are known, are given in table 5. From this it is evident that species from different sections have been examined, and that in all cases the chromosome number 2n = 30 has been established, except for a few populations of *Origanum vulgare* ssp. *vulgare*. All species investigated must be considered as diploids with a basic number x = 15. There are no reasons to doubt the validity of the aberrant number 2n = 32, which most probably is due to tetrasomatic aneuploidy. The chromosome number 2n = 30, found in the *Origanum* species, is not uncommon in related genera. Jalas & Kaleva (1967) reported it for some species of *Thymus* as well, and most *Micromeria* species so far investigated also have 2n = 30 (e.g. Strid, 1965 and Borgen, 1970).

I.9. CHEMICAL CHARACTERS

Hegnauer (1973) reports the occurrence of volatile oils, triterpene acids, phenols, sugars, and fatty acids in several *Origanum* species. He gave the following details. In *O. dictamnus* were extracted from the leaves a volatile oil containing 65 – 85% pulegon, and small amounts of ursol acid and oleanol acid. In *O. majorana* have been found in the leaves a volatile oil mainly containing terpines and small quantities of the acids just mentioned. In the seeds of this species have been detected

Table 5. Some data concerning chromosome numbers of *Origanum* species. The figure n = 15 for *O. dictamnus*, which was also reported by Lepper is omitted because it is doubtful whether the specimen in question was correctly identified, or the origin rightly given.

TAXON	SECTION	CHROMOSOME NUMBER	AUTHOR + DATE	REPORTED UNDER NAME
O. amanum	*Longitubus*	2n = 30	LEPPER, 1970	*O. amanum*
O. calcaratum	*Amaracus*	2n = 30	VON BOTHMER, 1970	*Amaracus tournefortii*
O. onites	*Majorana*	2n = 30	VON BOTHMER, 1970	*Majorana onites*
		2n = 30	MIÈGE & GREUTER, 1973	*O. onites*
O. vulgare ssp. *hirtum*	*Origanum*	2n = 30	VON BOTHMER, 1970	*O. heracleoticum*
O. vulgare ssp. *virens*	*Origanum*	2n = 30	LARSEN, 1960	*O. vulgare* var. *virens*
O. vulgare ssp. *vulgare*	*Origanum*	2n = 30	RUTLAND, 1941	*O. vulgare*
		2n = 30	GADELLA & KLIPHUIS, 1963	*O. vulgare*
		2n = 30	MÁJOVSKÝ, 1970	*O. vulgare* ssp. *vulgare*
		2n = 30	MARKOWA & IVANOVA, 1971	*O. vulgare* var. *vulgare*
		2n = 30	SKALIŃSKA & AL., 1971	*O. vulgare*
		2n = 32	SCHEERER, 1940	*O. vulgare*
		n = 16	HSU, 1968	*O. vulgare* var. *formosana*

planteose and some fatty acids. An analysis of the volatile oils of *O. syriacum* showed that it was rich in carvacrol and thymol. The same can be said of the volatile oil of *O. onites* (carvacrol and thymol together forming ca. 65 %). Thymian-like oils with high thymol and carvacrol contents have also been found in *O. floribundum* and *O. vulgare* ssp. *hirtum*, ssp. *virens* and ssp. *vulgare*. In addition ursol acid, oleanol acid and coffee acid have been isolated from leaves of *O. vulgare* ssp. *vulgare*, and stachyose from the roots.

A survey of rough analyses of volatile oils of several *Origanum* species has been given by Holmes (1913a, 1913b). For some species detailed analyses have recently been made of the essential oils, including analysis of *O. majorana* (e.g. Abou-Zied, 1973) and some subspecies of *O. vulgare* (e.g. Staicov, Zolotovitch & Kalaidjiev, 1968).

The most detailed study to date has been carried out by Maarse (1971), who worked with a commercially obtained strain of *O. vulgare* ssp. *vulgare*, the plants of which were grown in a botanic garden near Groningen (the Netherlands). For his analyses he mainly used a gaschromatograph, and additionally a column chromatograph and some kinds of spectrometers. His results can be summarized as follows. The essential oils contain 3 main fractions: the first chiefly contains monoterpene hydrocarbons, the second primarily monoterpenes oxygenated compounds and the third mainly sesquiterpene oxygenated compounds. From these main fractions he could identify 46 components in detail. Some 6 more were found, which could not be identified with certainty. The amount of volatile oil in the leaves increased early in the season, but decreased subsequently. The point of inversion was reached shortly before the flower buds were formed. A conclusive explanation of this phenomenon could not be found. Possibly the volatile oil is converted to less volatile components. It became clear, however, that evaporation did not cause the decrease of the volatile oil.

None of the chemical data mentioned have been used as criteria for delimitation of *Origanum*, its sections or species. The first reason for this is that the data are too fragmentary. Secondly many authors gave no or incomplete and inaccurate data about morphology, geography and taxonomy of the plants used (e.g. Calzolari, Stancher & Marletta, 1968).

I.10. GEOGRAPHY AND ECOLOGY

The species in the genus *Origanum* are mainly found in the Mediterranean region, as it is defined by, amongst others, Zohary (1973). Most species, viz. c. 75% of all, occur exclusively in the East Mediterranean subregion. The distribution of the *Origanum* species here agrees with the general plant geographical pattern in the Aegean (Runemark, 1971), as well as in Turkey (Davis, 1971). Some species of the section *Brevifilamentum* can be found in the adjacent parts of the Pontic province. Two species from the section *Campanulaticalyx* grow in the northeastern corner of the Saharo-Arabian region, near the border with the East Mediterranean subregion. The species in the section *Elongatispica* and one species from the section *Prolaticorolla* grow in the West Mediterranean subregion. Generally it must be remarked that several *Origanum* species grow nearly on or just to one side of both the West and East Mediterranean subregion borders. Partly this has to do with the fact that several species are Oro-Mediterranean elements. *O. vulgare* (section *Origanum*) possesses by far the largest distribution of all species in the genus. It is found throughout the Mediterranean region, but also in most parts of the Euro-Siberian region and in the Irano-Turanian region. For the moment it not quite clear whether the occurrence of *O. vulgare* in its most western (Azores, Canary Islands and Madeira) and most eastern sites (China, Taiwan) is a natural one or is due to introduction by man. For the occurrence of *O. vulgare* in North America the latter is certainly the case. C. 70% of all *Origanum* species must be considered as endemic: confined to one island or mountain (group). In figure 1, a distribution map is given for the genus. The distribution of each species is given in some further maps, and is treated in chapter II.

Not very much is known about the ecology of *Origanum* species. Zohary (l.c.) gives some details for a number of species. A few more data are available from herbarium labels. With regard to altitude, it can be said that many species in the

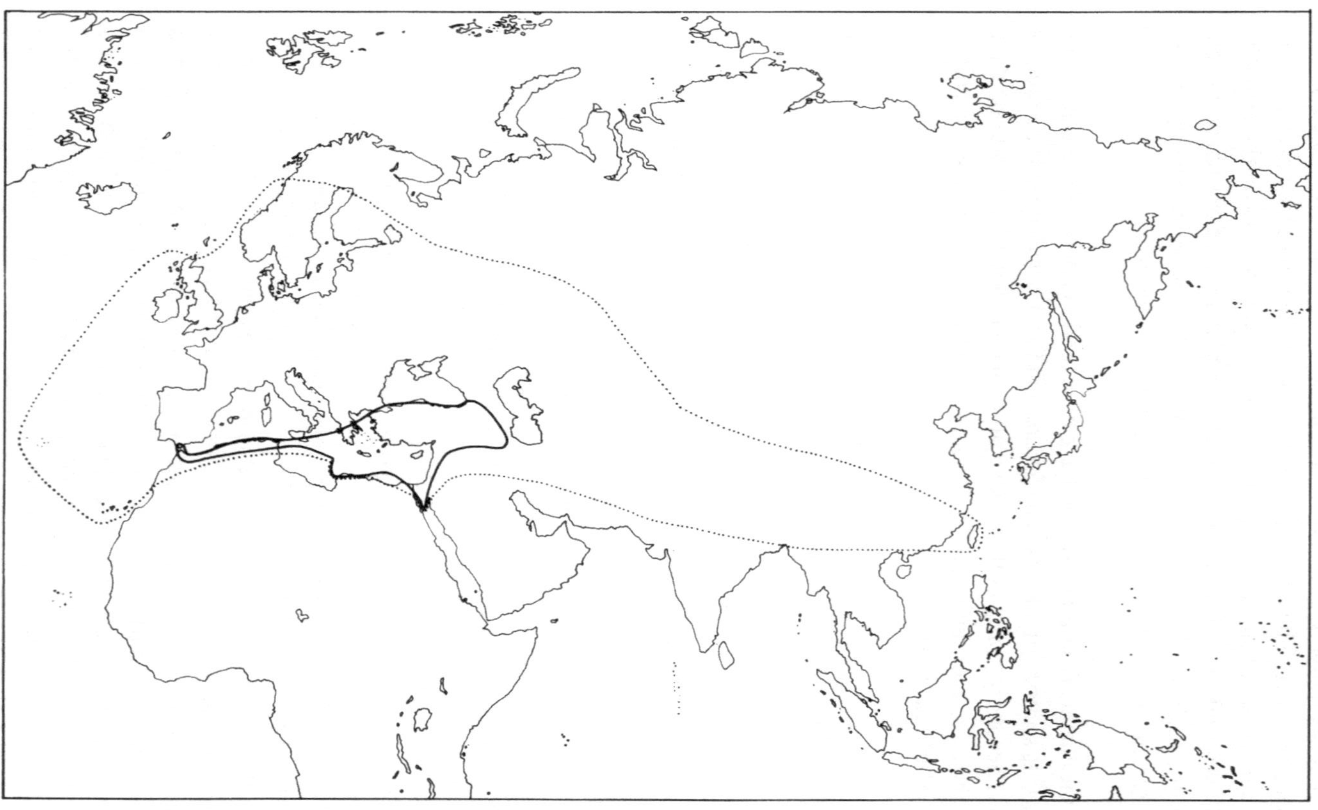

Figure 1. Distribution area of the genus *Origanum:* ....; area of all sections, except the section *Origanum:* ———.

genus are true to the generic name *Origanum*, which means "ornament of the mountains" (from the Greek words oros, mountain and hill, and ganos, ornament). Most species are found in headland and mountain areas from (0 –) 400 – 1800 (– 4000) metres. Many occur most abundantly at 1200 – 1500 metres, where they often grow in open coniferous (or mixed) woods in partial shade. A good example of such a species is *O. scabrum*, for which some ecological data have been given by Ietswaart, Fokkinga & Vroman (1972). However, some species, especially those in the section *Majorana*, often grow at a lower level. Zohary (l.c.) mentioned the occurrence of *O. syriacum* in several maquis, garigue and batha associations. *O. vulgare* is also often found at low altitudes. On the other hand it can grow as high as 4000 m, which further can only be said of *O. acutidens*. Most *Origanum* species grow exclusively on calcareous soils. The following, however, have been reported from non-lime substrates as well: *O. acutidens*, *O. dayi*, *O. elongatum* and *O. vulgare*. Krause, Ludwig & Seidel (1963) made analyses of vegetations on serpentine substrates on the island of Euboea, in which they mentioned *O. vulgare* and *O. scabrum*. Nearly all *Origanum* species grow on stony slopes and in rocky places, some are even strictly limited to cliffs. These species can be called chasmophytes. Snogrup (1971) gives a good example in *O. calcaratum;* other species of this type are *O. vetteri* and *O. boissieri*. In at least two syntaxa the name *Origanum* occurs. Zohary named one of the many associations he described for Palestine *Origanetum dayi*. In subcontinental Europe the order *Origanetalia vulgaris* is recognized by several authors, e.g. Westhoff & Den Held (1969).

I.11. HYBRIDIZATION

Some general remarks will be made here about hybridization in connection with the genus *Origanum*. The hybrids themselves will be treated individually after the species (chapter II.5). In 1895 Briquet mentioned a hybrid between species, belonging to what in this revision is understood as *Origanum*. Within the *Labiatae* as a whole hybridization was then considered to be a rare phenomenon. Since that time several other *Origanum* hybrids have become known and so Davis (1949) listed 7. An up to date survey in this respect for the genus is given in table 6, where 16 artificial and natural hybrids are listed. Those of the latter category have generally been found in small quantities. An example of the contrary, however, is *O.* × *lirium*, which forms a population on Euboea with a fairly normal reproduction.

The hybrids are in habit often somewhat like one of the parents. Usually they differ from both in size and shape and often colour of bracts, calyces and corollas. In these characters they are mostly intermediate between the parents. With respect to the fertility of the hybrids, stamens can be poorly or normally developed. In the first case no fertile pollen at all is produced, in the latter case usually only very small quantities of normal looking pollen are formed. No observations could be made on the development and fertility of seeds, and there are no references to these aspects in

SECTION	SECTION →	*Amaracus*		*Anatolicon*			*Brevifilamentum*	*Longitubus*	*Chilocalyx*		*Majorana*			*Origanum*			*Prolaticorolla*	
	SPECIES ETC.	*O. calcaratum*	*O. dictamnus*	*O. libanoticum*	*O. scabrum*	*O. sipyleum*	*O. bargyli*	*O. amanum*	*O. micranthum*	*O. microphyllum*	*O. majorana*	*O. onites*	*O. syriacum* var. *bevanii*	*O. vulgare* ssp. *hirtum*	*O. vulgare* ssp. *virens*	*O. vulgare* ssp. *vulgare*	*O. ehrenbergii*	*O. laevigatum*
Amaracus	*O. calcaratum*	\	X 3															
	O. dictamnus	X 3	\			X 4		X 3										
Anatolicon	*O. libanoticum*			\									X 1					
	O. scabrum				\									X 1				
	O. sipyleum		X 4			\						X 1		X 2				
Brevifilamentum	*O. bargyli*						\						X 1					
Longitubus	*O. amanum*		X 3					\										X 1
Chilocalyx	*O. micranthum*								\					X 2				
	O. microphyllum									\				X 1				
Majorana	*O. majorana*										\				X 4	X 4		
	O. onites					X 1						\		X 1				
	O. syriacum var. *bevanii*			X 1			X 1						\				X 1	X 1
Origanum	*O. vulgare* ssp. *hirtum*				X 1	X 2			X 2	X 1		X 1		\				
	O. vulgare ssp. *virens*										X 4				\			
	O. vulgare ssp. *vulgare*										X 4					\		
Prolaticorolla	*O. ehrenbergii*												X 1				\	
	O. laevigatum							X 1					X 1					\

Table 6. Origanum taxa, on the species level and below, between which hybrids are known. Each hybrid is represented by 2 x-marks. The following types are recognised:

1. hybrids found between the parents in natural surroundings,
2. putative hybrids from natural sites,
3. artificially made hybrids with cultivated parents,
4. hybrids with at least one parent cultivated.

the literature. It seems probable that no viable seeds are produced in most hybrids. From table 6 it is obvious that hybrids are possible between species from different sections and that no real barriers exist against hybridization throughout the genus. So hybrids may be invariably expected where two species are growing together in nature as well as in gardens. Summarizing it can be said that hybridization is a common phenomenon in the genus *Origanum* and that most hybrids have been found in small numbers, but a few in rather large numbers. So it can be concluded that the mixability of *Origanum* species is much less than the ability to cross.

Not only many hybrids are known in *Origanum* but also in the related genus *Thymus* (Jalas, 1972). Other genera in the *Melissinae/Thyminae* have not recently been the subject of extensive study. When such studies are carried out, most probably many more examples of hybridization will be found. The question arises whether these hybrids can only be found between species within a genus, or also between species of related, but different genera. Some odd "*Origanum*" forms with 2-lipped calyces and in some cases vary lax spikes, found as herbarium specimens, do to some extent support the hypothesis that hybridization also occurs beyond the generic limits in the *Melissinae/Thyminae*. Much investigation, in the field of biosystematics, is necessary to prove this hypothesis.

I.12. A HYPOTHESIS FOR SPECIATION

The following account has been developed for the origin of the species in *Origanum*, which must be seen as no more than an acceptable hypothesis. Other equally acceptable hypotheses could be made. It is probable that no one hypothesis exactly can account for what has happened in the past.

Ancestors of many species in *Saturejeae* genera spread in the Pliocene to Turkey and adjacent areas from the Irano-Turanean region. Partially through understood (and commonly accepted for all other groups as well) and partly through unknown mechanisms many new species originated. These species were for some time in balance with all climatological and ecological factors. Most species in the sections *Amaracus*, *Majorana* and also *O. vulgare* p.p. date from this time. So the ancestral forms in the genus could be described morphologically as follows. *Amaracus:* flowers in spikes, bracts large and purple, calyces 1-lipped for c. 3/4 with teeth in upper and lower lip (±) absent, corollas saccate; *Majorana:* flowers in dense spikes, bracts small and green, calyces 1-lipped for c. 9/10 with teeth in upper and lower lips absent, corollas flattened; *Origanum:* flowers in ± dense spikes, bracts small and green, calyces (sub)regularly 5-toothed, corollas of the common type. The *Amaracus* species then grew in hills at rather low altitudes. The other *Origanum* species were specialists of somewhat drier habitats near sea level.

In the late Pliocene and in the Pleistocene several changes in the climate occurred. In more recent times man added much to the changes by cutting down nearly all the woods (Zohary, 1973). Generally speaking the changes resulted in more arid conditions. The *Origanum* species were forced to move more or less into the mountains, where they came into contact with each other and with related species from the *Saturejeae*. Many chances arose for hybridization. Some of the original species lost only a part of their gene pool in hybrid populations and succeeded in maintaining themselves. Other original species will have dissolved completely in hybrid populations. A number of the hybrids so formed became equilibrated species. Most species in the sections *Anatolicon*, *Brevifilamentum*, *Chilocalyx*, *Elongatispica*, *Longitubus* and *Prolaticorolla* originated in this way.

To support the hypothesis given above some evidence will now be furnished. It is already clear that hybridization is a common phenomenon within the genera *Origanum* and *Thymus*. It is true that several *Origanum* hybrids occur in small numbers only, but *O.* × *lirium* is an example of a hybrid on its way to becoming species. Other examples of hybrids occurring in larger numbers are those mentioned in chapter II.5 under 2 and 15. Another argument is that several *Origanum* species are found (very) locally, e.g. *O. amanum*, *O. bilgeri*, *O. brevidens*, *O. ehrenbergii*, *O. grosii* and *O. hypericifolium*. The main arguments, however, are derived from the morphological characters of the *Origanum* species and hybrids. *O. hypericifolium* (sect. *Anatolicon*) is intermediate between the sections *Amaracus* and *Origanum* in its indumentum, size of bracts and flowers, as well as in the shape of the calyces. Its corolla is sometimes very slightly saccate. *O. hypericifolium* is somewhat like the hybrid *O. sipyleum* (sect. *Anatolicon*) × *O. vulgare* ssp. *hirtum* (sect. *Origanum*). Another species in the section *Anatolicon*, *O. vetteri*, is morphologically aberrant with respect to the pattern in the genus in its caespitose habit, in its almost unbranched stems, in its slightly revolute leaf margins. In the characters mentioned this species is more like a *Thymus* species. From *Thymus*, however it differs in among other things its less acute calyx lower lip teeth.

In the sect. *Brevifilamentum* most species possess more than 2 shortly pedunculate flowers per verticillaster, clearly 2-lipped calyces, and 2 short and 2 long stamens. These characters are not found in the section *Amaracus*, but they are present in the genus *Satureja* s.l. In habit and inflorescence the species in the section *Brevifilamentum* are much like those in the section *Amaracus*. So characters of the sect. *Amaracus* and the genus *Satureja* are both found in the sect. *Brevifilamentum*.

The same can be said about *O. amanum* (sect. *Longitubus*), which has long corollas with lips at right angles to the tube and nearly sessile stamens, characters found nowhere else in *Origanum*. In *Satureja* s.l. those large flowers are found in the "group" *Calamintha*.

In the section *Chilocalyx* a type of calyx is found resembling those of hybrids between *Majorana* species on one hand and *Origanum* species on the other (e.g. *O.* × *majoricum*). The small flowers of some species in this section could be derived from *Satureja* s.l. species in the "group" *Micromeria*. So it seems plausible to suppose characters of the sections *Majorana* and *Origanum* and/or *Satureja* are present in the section *Chilocalyx* because of hybridization. Influences from *Satureja* in the section *Prolaticorolla* could be the following: corollas with longer tubes, short staminal filaments, the rather frequent occurrence of ± lax spikes in *O. laevigatum* and *O. ehrenbergii*, and very shortly branched inflorescences in *O. compactum*.

The very lax spikes of the species in the section *Elongatispica* together with the small flowers could have come from the "group" *Micromeria* into *Origanum* through hybridization with *O. vulgare* ssp. *hirtum*. There are indications that some of the processes of hybridization took place rather recently, and possibly are due to man's intervention in nature.

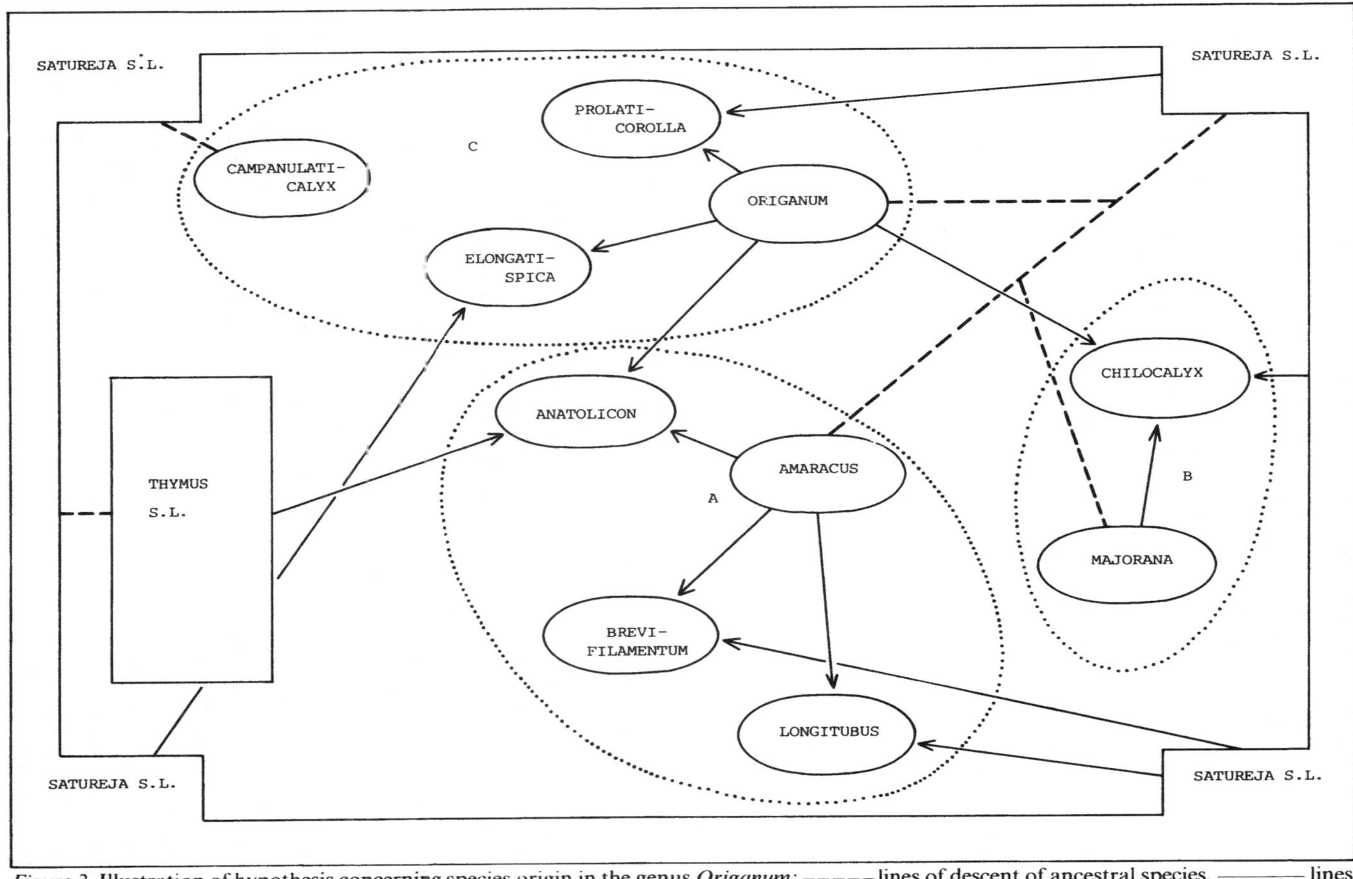

Figure 2. Illustration of hypothesis concerning species origin in the genus *Origanum:* – – – – – lines of descent of ancestral species, ———— lines of genesis of secundary species through hybridization. A, B and C are groups of related sections.

In the hypothesis mentioned hybrization must be considered as the most important speciation mechanism in the genus *Origanum*. A relatively slight contribution to speciation in the genus can be attributed to geographical differentiation, namely within *O. vulgare*. Here the ssp. *hirtum* must be considered as the oldest one. The other ssp. could have been developed from this through geographical differentiation. For the moment there is no reason for presuming that speciation in the genus took place through polyploidization.

As for the section *Campanulaticalyx*, it can be remarked that here probably a partly parallel development is involved, originating from another part of *Satureja* s.l., than from which the greater part of *Origanum* is derived.

In figure 2 some aspects of the hypothesis mentioned above are illustrated. In this figure *Satureja* s.l. stands for the genera *Acinos*, *Calamintha*, *Clinopodium*, *Micromeria* and *Satureja* s.s., as mentioned in subchapter I.5, and their ancestors.

I.13. CULTIVATION OF ORIGANUM SPECIES

There are four ways in which *Origanum* species are used by man: as medicinal herbs, as culinary herbs, as garden plants and also as a dye.

Origanum species have been used as medicinal herbs from very early times. The "hyssop" of the bible is possibly *O. syriacum (O. maru)*, at least partly, as is shown by Andrews (1961). This author gives an interesting survey of the use of several *Labiatae* in the East Mediterranean area in ancient times, from which it is also clear that the use of these plants as culinary herbs played a very minor role at that time. The island of Kriti was an important export and transit country for these herbs in the beginning of our era. Fraas (1845) gives seven *Origanum* names in his survey of classic plants with several synonyms of Dioscorides, Hippocrates, Theophrastus and Plinius. As an example of a prelinnean herbal is chosen here "De Nederlandtse Herbarius of Kruydt-boeck" of Nylandt (1682). In this book are treated three "species of *Majorana*" and one of "*Origanum*", representing probably two *Origanum* species according to the present revision. For each "species" a description is given, also the healing power and instructions for cultivation, and references are made to the classic botanists just mentioned.

Kosteletzky (1831) gives some 15 *Origanum* "species" and their medicinal uses. The work ("Algemeine Medizinisch-Pharmaceutische Flora") is a good example of the species nomenclature that was used in those days: of these 15 "species" only four are recognized in this revision. Nowadays *Origanum* is only rarely used as a medicinal herb (Braun, 1974).

Several works do not treat the use of *Origanum* species as medicinal herbs separately from their use as culinary herbs. A well-known example is Miller (1768). In the 8th edition of his "Gardeners Dictionary" he lists no less than 13 *Origanum* "species" with descriptions, growing instructions and various uses. Among more recent works is that of Grieve (1931). This work gives four "good" *Origanum* species, all under the English name Marjoram. This introduces the question as to which species can be found under common names like: Marjoram, Marjolein, Marjolaine, Majoran, Origan and Dost, sold as herbs. Generally speaking all these herbs can be identified from some calyces, bracts and leaves. It should be kept in mind that mixtures of more than one species can occur. It is also quite possible that species other than *Origanum* are recognized under the names mentioned. This has been made clear by Calpouzos (1954), who analysed many samples from the American market, and found beside *Origanum* species also *Lippia graveolens* and more seldom *L. palmeri*, which belong to the *Verbenaceae*. Most of the samples containing the last two species came from Mexico. From a study of the literature he found many more species described as Oregano, belonging to the following genera and families: *Coleosanthus (Compositae)*, *Calamintha*, *Coleus*, *Hedeoma*, *Hyptis*, *Monarda*, *Ocimum*, *Poliomintha*, *Satureja* (all *Labiatae*), *Borreria (Rubiaceae)*, *Limnophila (Scrophulariaceae)*, *Eryngium (Umbelliferae)* and *Lantana (Verbenaceae)*. In Europe the following *Origanum* species (forms) can be expected when buying herbs called Oregano, Marjoram etc.: *O. majorana*, *O. onites*, *O. syriacum*, *O. vulgare* and hybrid forms between the latter species and *O. majorana*. In the market of Herakleon (Kriti) I observed sale of *O. dictamnus*, *O. microphyllum*, *O. onites* and *O. vulgare* ssp. *hirtum*. Near Anavrouti (Taygetos Mts., southern Greece) *O. scabrum* was used as a herb. Holmes (1913b) mentioned as (potential) suppliers of the Marjoram oil: *O. compactum*, *O. floribundum*, *O. majorana*, *O. onites*, *O. syriacum*, *O. vulgare* ssp. *hirtum* and *O. vulgare* ssp. *vulgare* (translated in the nomenclature of this revision). He gave some morphological details of these species as well as some chemical and physical characters of the different types of oil. Chevalier (1938) gives many data on the cultivation of *O. majorana* (and some other *Origanum* species) in France and North Africa. Gams (1927) gives some details in this respect for Central Europe together with interesting remarks on the (history of) the uses of the *Origanum* herbs, which will not be discussed here.

As ornamentals in the garden the following *Origanum* species can be found: *O. amanum*, *O. calcaratum* (= *O. tournefortii*), *O. dictamnus*, *O.* × *hybridinum (O. pulchellum)*, *O. laevigatum*, *O. leptocladum*, *O. libanoticum*, *O. majorana*, *O. microphyllum*, *O. onites*, *O. scabrum* (= *O. pulchrum*), *O. rotundifolium*, *O. syriacum* (= *O. maru*) and *O. vulgare*. Of the latter species two cultivars are known: '*Aureum*' with golden-yellow leaves and '*Humile*', a dwarf form, no more than 20 cm high. All species mentioned, except *O. vulgare*, are best grown in a well drained, limy soil, in a sunny position preferably in a rock garden or under alpine house conditions. In West Europe they need some winter protection. More details on the cultivation are given by Elliot (1966) for England, and by Wolf (1954) for North America.

In one case the use of *O. vulgare* as a dye is mentioned: the sap gives a red colouring (Chambert, 1895).

I.14. PHYTOPARASITES

Two rust fungi have been reported on *O. vulgare*. One, *Puccinia rübsaameni*, is described exclusively from this species (Gäumann, 1959). The other, *P. menthae*, has been found on 55 European *Labiatae* from 12 genera (also reported by Gäumann) and on 25 American *Labiatae* from 8 genera (Baxter & Cummins, 1953). Most of these genera belong to the tribe of the *Saturejeae*. Mühle (1956) suggests that *P. menthae* also may occur on *O. majorana*.

Schmelzer (1963) mentioned the occurrence of a strain of alfalfa mosaic virus on commercially grown specimens of *O. majorana*, in fields near Aschersleben (D.D.R.).

Miličić & Plakolli (1974) reported the same virus infection in *O. vulgare* grown in the Botanical Garden of Zagreb.

I.15. REFERENCES

ABOU-ZIED, E.N. 1973. The seasonal variations of growth and volatile oil in the two introduced types of Majorana hortensis Mnch., grown in Egypt. Pharmazie 28: 55 – 56.

ANDREWS, A.C. 1961. Hyssop in the classical era. Class. Philology 56: 230 – 248.

BAXTER, J.W. & G.B. CUMMINS. 1953. Physiologic specialization in Puccinia menthae Pers., and notes on epiphytology. Phytopathology 43: 178 – 180.

BENTHAM, G. 1934. Labiatarum Genera et Species etc.: 333 – 339. Ridgway, London.

— 1836. Labiatae orientales etc. Ann. Sc. Nat. 2(6): 43.

— 1848. In: de Candolle, Prodromus Systematis Naturalis etc. 12: 191 – 197. Masson, Paris.

— & J.D. HOOKER. 1876. Genera plantarum 2: 1185 – 1186. Reeve & Co., Williams & Norgate, London.

BOISSIER, E. 1844. Diagnoses Plantarum Orientalium Novarum 1(5): 14 – 15. Hermann, Leipzig.

— 1846. Diagnoses Plantarum Orientalium Novarum 1(7): 48. Hermann, Leipzig.

— 1854. Plantes nouvelles etc. Ann. Sc. Nat. 4(2): 252 – 253.

— 1859. Diagnoses Plantarum Orientalium Novarum 2(4): 9 – 12. Hermann, Leipzig, Baillière, Paris.

— 1879. Flora Orientalis 4: 546 – 554. George, Genève, Basel.

BORGEN, L. 1970. Chromosome numbers of Macaronesian flowering plants. Mag. Bot. 17: 145 – 161.

BORNMÜLLER, J. 1898. Ein Beitrag zur Kenntniss der Flora von Syrien und Palästina. Verh. Zool.-Bot. Ges. Wien 48: 615 – 616.

— 1917. Zur Flora des nördlichen Syriens. Notizbl. Bot. Gart. Mus. Berlin-Dahlem 7(63): 26 – 28.

BOTHMER, R. VON. 1970. Studies in the Aegean Flora XV, Chromosome Nùmbers in Labiatae. Bot. Not. 123: 52 – 60.

BRAUN, H. 1974. Heilpflanzen-Lexikon für Ärzte und Apotheker: 116, 133 – 134. Fischer, Stuttgart.

BRIQUET, J. 1895. Labiatae. In: Engler & Prantl, Natürlichen Pflanzenfamilien etc. 4(3a): 183 – 384. Engelmann, Leipzig.

CALPOUZOS, L. 1954. Botanical aspects of Oregano. Econ. Bot. 8: 222 – 233.

CALZOLARI, D., B. STANCHER & G. MARLETTA. 1968. Origanum oils and their investigation by gas-chromatographic and infrared spectroscopic analysis. Analyst 93: 311 – 318.

CHAMBERT. A. 1895. De l'emploi populaire des plantes sauvages en Savoie. Bull. Herb. Boiss. 3: 336.

CHEVALIER, A. 1938. La Marjolaine vraie (Majorana hortensis) et sa culture. Revue Bot. Appliqué etc. 18: 593 – 604.

DANIN, A. 1967. A new Origanum from Israel. Israel J. Bot. 16: 101 – 103.

— 1969. A new Origanum from the Isthmic Desert (Sinai). Israel J. Bot. 18: 191 – 193.

DAVIS, P.H. 1949. Additamenta ad floram anatoliae I. Kew Bull. 1949: 404 - 411.

— 1951. Additamenta ad floram anatoliae II. Kew Bull. 1951: 86 - 89.

— 1956. Notes on the summer flora of the Aegean. Notes Roy. Bot. Gard. Edinburgh 21: 137 - 138.

— 1971. Distribution patterns in Anatolia with particular reference to endemism. In: Davis, Harper & Hedge, Plant Life of South-West Asia: 15 - 27. The Botanical Society, Edinburgh.

EL-GAZZAR, A. & L. WATSON. 1970. A taxonomic Study of Labiatae and Related Genera. New Phytol. 69: 451 - 486.

ELLIOTT, R.C. 1966. Of Marjorams and Dittanies. Quart. Bull. Alpine Gard. Soc. 34: 198 - 205.

FERNANDES, R. & V.H. HEYWOOD. 1972. Origanum. In: Tutin et al., Flora Europaea 3: 171 - 172. University Press, Cambridge.

FRAAS, C. 1845. Synopsis Plantarum Florae Classicae etc.: 181 - 182. Fleischmann, München.

GADELLA, TH.W.J. & E. KLIPHUIS. 1963. Chromosome numbers of flowering plants in the Netherlands. Acta Bot. Neerl. 12: 195 - 230.

GAMS, H. 1927. Labiatae. In: Hegi, Illustrierte Flora von Mittel-Europa etc. 5: 2327 - 2335. Hanser, München.

GÄUMANN, E. 1959. Die Rostpilze Mitteleuropas. In: Beiträge zur Kryptogamenflora der Schweiz 12: 1002 - 1012.

GLEDITSCH, J.G. 1764. Systema Plantarum a Staminum Situ etc.: 189. Haude & Spener, Berlin.

GRIEVE, M. 1931. A Modern Herbal: 147, 519 - 521. Cape. London.

HALÁCSY, E. VON. 1902. Conspectus Florae Graecae 2: 552 - 557. Engelmann, Leipzig.

HAYEK, A. 1931. Prodromus Florae Peninsulae Balcanicae 2 (edendum et curavit Fr. Markgraf). Rep. Spec. Nov. 30: 332 - 336.

HEGNAUER, R. 1973. Chemotaxonomie der Pflanzen 4. Birkhäuser, Basel, Stuttgart.

HOFFMANNSEGG, J.C. & H.F. LINK. 1809. Flora Portugaise etc. 1: 119 - 121. Berlin.

HOLMES, E.M. 1913a. The Oils of Marjoram of Commerce II. Perf. Ess. Oil Rec. 4: 41.

— 1913b. The Oils of Marjoram of Commerce III. Perf. Ess. Oil Rec. 4: 69 - 75.

HSU, CH. 1968. Preliminary chromosome studies on the vascular plants of Taiwan (II). Taiwania 14: 11 - 27.

IETSWAART, J.H. 1980. Materials for a Flora of Turkey. Origanum. Not. R. B. G. Edinburgh 38(1): 46 - 47.

— 1981. Origanum. In: Davis, Flora of Turkey and the East Aegean Islands 7 (in press).

—, A. FOKKINGA & M. VROMAN. 1972. Delimitation of Origanum scabrum Boiss. et Heldr. (Labiatae) by means of morphological criteria. Acta Bot. Neerl. 21: 439 - 447.

JALAS, J. 1972. Thymus. In: Tutin et al., Flora Europaea 3: 172 - 182. University Press., Cambridge.

— & K. KALEVA. 1967. Chromosome studies in Thymus L. (Labiatae) V. Ann. Bot. Fennici 4: 74 - 80.

KHEYR-POUR, M.A. 1969. Contribution à l'étude du déterminisme génétique et écologique du polymorphisme sexuel chez Origanum vulgare L. Compt. Rendus Séances Acad. Sc. 268, Série D: 2566 - 2567.

KOCH, K. 1848. Beiträge zu einer Flora des Orientes. Linnaea 21: 639 - 663.

KOSTELETZKY, V.F. 1834. Allgemeine Medizinische-Pharmazeutische Flora etc. 3: 767 - 770. Borrosch & André, Praha.

KRAUSE, W., W. LUDWIG & F. SEIDEL. 1963. Zur Kenntnis der Flora und Vegetation auf Serpentinstandorten des Balkans 6. Bot. Jahrb. 82: 337 - 403.

KUNTZE, O. 1867. Taschenflora von Leipzig: 106. Winter, Leipzig, Heidelberg.

— 1891. Revisio Generum Plantarum etc. 2: 528. Felix, Leipzig.

LAMARCK, J.B.A.P.M. DE. 1798. Encyclopédie Méthodique. Botanique 4: 605 - 609. Agasse, Paris.

LARSEN, K. 1960. Cytological and experimental studies on the flowering plants of the Canary Islands. Kong. Danske Vidensk. Selskab. Biol. Skrifter 11: 1 - 62.

LEPPER, L. 1970. Beiträge zur Chromosomenzahlen-Dokumentation. Wiss. Zeitschrift 19: 369 - 376.

LEWIS, D. & L.K. CROWE. 1956. The genetics and evolution of gynodioecy. Evolution 10: 115 - 125.

LINK, H.F. 1822. Enumeratio Plantarum Horti Regii Botanici Berolinensis Altera 2: 113 - 115. Reimer, Berlin.

LINNAEUS, C. 1753. Species Plantarum, facsimile edition 1959, 2: 588 - 590. Quaritch, London.

— 1754. Genera Plantarum, facsimile edition 1960: 256. Engelmann (Cramer), Weinheim, Wheldon & Wesley, Codicote.

— 1763. Species Plantarum. Edition II, 2: 822 - 825. Salvius, Stockholm.

MAARSE, H. 1971. Samenstelling van de vluchtige olie van Origanum vulgare L. ssp. vulgare gedurende de ontwikkeling van de plant. Thesis, Groningen.

MÁJOVSKY, J. 1970. Index of chromosome numbers of Slovakian flora 2. Acta Fac. Nat. Univ. Comenianae, Bot. 18: 45 – 56.

MARKOWA, M. & P. IVANOVA. 1971. Karyologische Untersuchungen der Vertreter der Boraginaceae, Labiatae and Scrophulariaceae in Bulgarien II. Mitt. Bot. Inst. Sofia 21: 123 – 131.

MELCHIOR, H. 1964. Engler's Syllabus der Pflanzenfamilien: 438 – 444. Borntraeger, Berlin-Nikolassee.

MIÈGE, J. & W. GREUTER. 1973. Nombres chromosomiques de quelques plantes récoltées en Crète. Ann. Mus. Goulandris 1: 105 – 111.

MILIČIĆ, D. & M. PLAKOLLI. 1974. Spontaneous infection of some Labiates with alfalfa mosaic virus. Acta Bot. Croatica 33: 9 – 15.

MILLER, P. 1754. The Gardeners Dictionary. Abridged Edition IV, facsimile edition 1969: 829 – 830, 982 – 983, Cramer, Lehre.

— 1768. Gardeners Dictionary etc. Edition VIII: pages concerning Origanum. Rivington etc., London.

MOENCH, C. 1794. Methodus Plantas Horti Botanici et Agri Marburgensis, a Staminum Situ Describendi: 405 – 406, Libraria Academiae, Marburg.

— 1802. Supplementum ad Methodum Plantas a Staminum Situ Describendi: 137. Libraria Academiae, Marburg.

MOUTERDE, P. 1935. Trois hybrides d'Origanum Maru L. Ann. Fac. Fr. Médicine Beyrouth 6: 303 – 311.

— 1973. Novitates florae libano-syriacae. Saussurea 4: 22 – 23.

MÜHLE, E. 1956. Die Krankheiten und Schädlinge der Arznei-, Gewürz- und Duftpflanzen. Deutsche Akad. Landwirtsch. Wissensch. Abh.: 17.

NYLANDT, P. 1682. De Nederlandtse Herbarius of Kruydt-boeck etc., facsimile edition 1976. 229 – 240. Interbook International, Schiedam.

NYMAN, C.F. 1881. Conspectus Florae Europaeae etc.: 592. Bohlin, Örebrö.

— 1881. Conspectus Florae Europaeae. Supplementum 2: 256. Bohlin, Örebrö.

POST, G.E. 1893. In: Post & Autran, Plantae Postianae. Bull. Herb. Boiss. 1: 405.

— 1895. In: Post & Autran, Plantae Postianae. Bull. Herb. Boiss. 3: 161 – 162.

— & J.E. DINSMORE. 1933. Flora of Syria, Palestine and Sinai 2: 332 – 335. American Press, Beirouth.

RAFINESQUE, C.S. 1936. Flora Telluriana 3: 86 – 87. Probasco, Philadelphia.

— 1840. Autikon Botanikon 1: 118 – 119. Philadelphia.

RECHINGER, K.H. fil. 1938. Majoranamaracus Zernyi etc. nov. hybr. Rep. Spec. Nov. 45: 95.

— 1943a. Flora Aegaea. Denkschr. Akad. Wiss. Wien, Math.-Nat. Kl. 105(1): 530 – 533.

— 1943b. Neue Beiträge zur Flora von Kreta. Denkschr. Akad. Wiss. Wien, Math.-Nat. Kl. 105(2): 125 – 127.

— 1952. Labiatae novae orientales. Österr. Bot. Zeitschr. 99: 63 – 64.

— 1961. Die Flora von Euboea. Bot. Jahrb. 80: 395 – 396.

— & L. EDELBERG. 1954. In: Køie & Rechinger, Symbolae Afghanicae I. Kong. Danske Vidensk. Selskab, Biol. Skrifter 8: 75 – 76.

RUNEMARK, H. 1971. Distribution patters in the Aegean. In: Davis, Harper & Hedge, Plant Life of South-West Asia: 3 – 14. The Botanical Society, Edinburgh.

—, S. SNOGERUP & B. NORDENSTAM. 1960. Studies in Aegean Flora 1, Floristic Notes. Bot. Notiser 113: 421 – 450.

RUTLAND, J.P. 1941. A list of chromosome numbers of British plants, Supplementum no. 1. New Phytol. 40: 211 – 213.

SALISBURY, R.A. 1796. Prodromus Stirpium in Horto ad Chapel Allerton Vigentium: 85. London.

SAVI, C.G. 1840. Osservazioni sopra algune Specie del Genere Origanum etc.: 1 – 20. Pisa.

SCHEELE, A. 1843. Beiträge zur deutschen und schweizerischen Flora. Flora (Allgem. Bot. Zeit.) 26: 573 – 575.

SCHEERER, H. 1940. Chromosomenzahlen aus der Schleswig-Holsteinischen Flora II. Planta 30: 716 – 725.

SCHMELZER, K. 1963. Das Luzernemosaik-Virus am Majoran (Majorana hortensis Moench). Nachrichtenbl. deutsche Pflanzenschutzdienst, Neue Folge 17: 108 – 111.

SIBTHORP, J. & J.E. SMITH. 1826. Florae Graeca 6: 56 – 59. London.

SKALIŃSKA, M., A. JANKUN & H. WCISŁO. 1971. Studies in chromosome numbers of Polish Angiosperms, eighth contribution. Acta Biol. Cracov. Ser. Bot. 14: 76 – 77.

SNOGERUP, S. 1971. Evolutionary and Plant Geographical Aspects of Chasmophytic Communities. In: Davis, Harper & Hedge, Plant Life of South-West Asia: 157 – 170. The Botanical Society, Edinburgh.

STAICOV, V., G. ZOLOTOVITCH & I. KALAIDJIEV. 1968. Origanum heracleoticum. L. Plant and Essential Oil. Soap, Perf. Cosmetic 41: 327 – 330.

STOKES, J. 1812. A Botanical Materia Medica etc. 3: 343 – 352. Johnson, London.

STRID, A. 1965. Studies in the Aegean flora VI, notes on some genera of Labiatae. Bot. Not. 118: 104 – 122.

THIEBAUT, J. 1953. Flore Libano-Syrienne 3: 45 – 47. Centre National de la Recherche Scientifique, Paris.

TUTIN, T.G., V.H. HEYWOOD, N.A. BURGES, D.M. MOORE, D.H. VALINTINE, S.M. WALTERS & D.A. WEBB. 1972. Flora Europaea 3: 126 – 192. University Press, Cambridge.

VALDEYRON, G., B. DOMMÉE & A. VALDEYRON. 1973. Gynodioecy: another computer simulation model. Amer. Naturalist 107: 454 – 459.

VOGEL, TH. 1840. In: Marquart & Vogel, Beiträge zur Geschichte der Herba Origani cretici. Buchners Repert. Pharm. 12: 289 – 311.

— 1841. Bemerkungen über einige Arten aus den Gattungen Thymus und Origanum. Linnaea 15: 74 – 82.

WATERMAN, A.H. 1960. Pollen grain studies of the Labiatae of Michigan. Webbia 15: 399 – 415.

WESTHOFF, V. & A.J. DEN HELD. 1969. Plantengemeenschappen in Nederland: 228 – 229. Thieme, Zutphen.

WILLDENOW, C.L. 1800. Caroli a Linné species plantarum etc. 3: 132 – 138. Nauk, Berlin.

WILLKOMM, H.M. & J.M.C. LANGE. 1868. Prodromus Florae Hispanicae etc. 2: 398 – 399. Schweizerbart, Stuttgart.

WOLF, G.P. DE. 1954. Notes on cultivated Labiates 3, Origanum and relatives. Baileya 2: 57 – 66.

WUNDERLICH, R. 1967. Ein Vorschlag zu einer natürlichen Gliederung der Labiaten auf Grund der Pollenkörner, der Samenentwicklung und des reifen Samens. Österr. Bot. Zeitschr. 114: 383 – 483.

ZOHARY, M. 1973. Geobotanical Foundations of the Middle East 1 & 2. Fischer, Stuttgart, Swets & Zeitlinger, Amsterdam.

II. TAXONOMIC TREATMENT

II.1. INTRODUCTION

All taxa are treated together (II.4), except the hybrids to which a separate subchapter is devoted (II.5). The general key (II.3) divides in 3 subkeys, each dealing with the species in related sections. Thirty eight species are recognized, one with 6 subspecies and another with 3 varieties, grouped in 10 sections. In addition 17 hybrids are recognized. Criteria are given in I.6 for what is understood by the terms section, species, subspecies and variety.

One or some type specimen(s) is (are) designated for each species, subspecies and variety. A type species is also designated for each section recognized in this revision. For the species, subspecies and varieties usually a holotype, and not seldom one or more isotypes were available for the correct and accepted name. In a number of cases (e.g. the Linnean types) only photographs were available for study. In some cases a neotype had to be chosen, because a holo-, iso- or lectotype are known not to exist.

In descriptions of species, subspecies and varieties many measurements are given, which in general are based on measurings of 10 – 15 specimens for each taxon. In a number of cases however only three specimens or less were available. The number of specimens measured can be deduced from the number of specimens cited as "specimens studied". For several characters of a specimen a mean, a maximum and a minimum value were established. The mean value for a species character is obtained by averaging the means of all specimens in question. The maximum (minimum) value given is the maximum (minimum) value for all specimens studied. The lengths of the leaves are always given excluding the petioles; those of the branches without the spikes, and those of the spikes without the protruding flowers. Most terms used in the descriptions are borrowed from Stearn's "Botanical Latin". Usually the descriptions of the taxa are based on plants from natural sites. In cultivated specimens stems and branches are often more numerous and longer, while their indumentum is less developed. The spikes are often less compact and the bracts less purple coloured. (This also should be kept in mind when using the keys.)

For all specimens from Turkey the new names of provinces (vilayets) are given, instead of the names which were often found on the herbarium labels.

In general one species is figured together with all details for each section. Furthermore the differentiating characters of each species are figured, for the species of one section together, in one or two figures. The geographical distribution of each species is given in another figure, for all species in one section together.

Of the synonyms only a small number of types could be obtained for verification, many were untraceable. All names which could not be ascertained as synonyms have been summed up as "nomina dubia" (II.6), while *Origanum* names which do not belong to the genus as conceived here, are given under "species excludenda" (II.7).

With the "specimens studied" the herbaria are given according to the Index Herbariorum, as follows:

AVU	Biologisch Laboratorium, Afdeling Plantensystematiek, Vrije Universiteit, Amsterdam
B	Botanisches Museum, Berlin
BM	British Museum (Natural History), London
BP	Museum of Natural History (Department of Botany), Budapest
—	Private herbarium of Dr. Buttler, München
C	Botanical Museum and Herbarium, Copenhagen
CAI	Department of Botany (Faculty of Sciences), Cairo
CAT	Istituto di Botanica, Orto Botanico, Catania
COI	Botanical Institute of the University, Coimbra
E	Royal Botanic Garden, Edinburgh
FI	Herbarium Universitatis Florentinae (Istituto Botanico), Firenze
G	Conservatoire et Jardin botaniques, Genève
—	Private herbarium Dr. Huber-Morath, Basel
HUJ	Department of Botany, Hebrew University, Jerusalem
JE	Institut für Spezielle Botanik und Herbarium Haussknecht, Jena
K	The Herbarium and Library, Kew
L	Rijksherbarium, Leiden
LINN	The Linnean Society, London
MA	Instituto "Antonio José Cavanilles", Jardin Botánico, Madrid
MPU	Institut de Botanique, Montpellier
OXF	Fielding Herbarium, Druce Herbarium (Department of Botany), Oxford
P	Muséum National d'Histoire Naturelle, Laboratoire de Phanérogamie, Paris
PAD	Istituto Orto Botanico dell'Università, Padova
PRC	Universitatis Carolinae, Facultatis Biologicae Scientiae Cathedra, Praha
—	Private herbarium of Dr. Sorger, Wien
S	Naturhistoriska Riksmuseum (Botanical Department), Stockholm
U	Instituut voor Systematische Plantkunde, Utrecht
W	Naturhistorisches Museum, Wien
WU	Botanisches Institut und Botanischer Garten der Universität, Wien

II.2. ORIGANUM

Origanum Linnaeus, Gen. Pl.: 256 (1754). – Type: *Origanum vulgare* Linnaeus, according to Britton & Brown (1913).

Majorana Miller, Gard. Dict., Abr. IV Ed.: 829 (1754), (nom. gen. cons.). – Type: *Majorana hortensis* Moench (= *Origanum majorana* Linnaeus).

Amaracus Hill, Br. Herb.: 381 (1756). – Type: *Amaracus vulgaris* Hill (= *Origanum majorana* Linnaeus).

Dictamnus Hill, Br. Herb.: 381 (1756). – Type: *Dictamnus creticus* Hill (= *Origanum dictamnus* Linnaeus).

Hofmannia Heister ex Fabricius, Enum. Meth. Pl.: 1 (1759), (nom. gen. rejec.). – Type: "*Dictamnus montis Sipyli*" (= *Origanum sipyleum* Linnaeus).

Amaracus Gleditsch, Syst. Pl.: 189 (1764), (nom. gen. cons.). – Type: *Amaracus tomentosus* Moench (= *Origanum dictamnus* Linnaeus).

Beltokon Rafinesque, Fl. Tell. 3: 86 (1836). – Type: *Beltokon tourneforti* (Aiton) Rafinesque (= *Origanum calcaratum* Jussieu).

Onitres Rafinesque, Fl. Tell. 3: 86 (1836). – Type: *Onites tomentosa* Rafinesque (= *Origanum onites* Linnaeus).

Oroga Rafinesque, Fl. Tell. 3: 86 (1836). – Type: *Onites hereaclontica* Rafinesque (= *Origanum vulgare* Linnaeus ssp. *hirtum* (Link) Ietswaart).

Schizocalyx Scheele, Flora, Neue Reihe 1: 575 (1843). – Type: *Schizocalyx syriacus* (Linnaeus) Scheele (= *Origanum syriacum* Linnaeus).

× *Origanomajorana* Domin, Preslia 13–15: 197 (1935). – Type: × *Origanomajorana applii* Domin (= *Origanum* × *applii* (Domin) Boros).

× *Majoranamaracus* Rechinger, Fedde Rep. 45: 95 (1938). – Type: × *Majoranamaracus zernyi* Rechinger (= *Origanum* × *adonidis* Mouterde).

Non *Zatarendia* Rafinesque, Fl. Tell. 3: 86 (1836). – Type: *Zatarendia aegyptiaca* (Linnaeus) Rafinesque (= *Origanum aegyptiacum* Linnaeus, species excludenda).

Subshrubs or perennial herbs, scabrous, hirtellous, hirsute, (appressed) pilose, pubescent, tomentose or lanate with multicellular, simple hairs (seldom hairs 1-celled or branched), or glabrous and often glaucous; stalked and sessile glands present, the latter often conspicuous in the leaves. Stems several, ascending or erect. Branches usually present in 1 – 3 orders. *Leaves* oval, ovate, heart-shaped or roundish, entire, serr(ul)ate or crenulate, tops obtuse to acuminate; (sub)sessile, or petiolate in the lower parts of the stems to subsessile in the upper parts; petioles up to c. $\frac{1}{2}$ as long as the blades. Verticillasters aggregated in dense, sometimes loose, spikes. *Spikes* often arranged in paniculate or corymbiform inflorescences, subglobose, ovoid, (quadrigonous-) cylindrical, seldom pyramid-shaped, erect or nodding. *Bracts* always distinct from the leaves in shape and size, often also in texture and colour, usually imbricate, $\frac{1}{2}$ – 3 times as long as the calyces, roundish, (ob)ovate or oval, tops obtuse to acuminate; when clearly longer than the calyces membranous and (partly) purple or yellowish green, when ± as long as the calyces ± leaf-like in texture and colour. *Flowers* bisexual or female, usually 2, sometimes several per verticillaster. *Calyces either* tube shaped, sometimes campanulate or turbinate, (sub) regularly 5-toothed for c. $\frac{1}{3}$, or 2-lipped or 1-lipped for $\frac{1}{5}$ – $\frac{3}{5}$, with c. 13 veins, *or* flattened and 1-lipped for c. $\frac{9}{10}$, with c. 10 veins; upper lips with 3 teeth or lobes or (sub)entire; lower lips, when present c. $\frac{1}{5}$ – 1 times as long as the upper lips, consisting of 2 teeth or lobes; throats usually with a hair ring. *Corollas* 2-lipped for $(\frac{1}{7})\frac{1}{5}$ – $\frac{2}{5}$, $1\frac{1}{2}$ – 3(4) times as long as the calyces, sometimes saccate or flattened, with 9 veins, purple, pink or white; upper lips emarginate or with 2 short lobes, lower lips ± as long as the upper lips, divided, for c. $\frac{1}{2}$, into 3 subequal or unequal lobes; the middle one longest. *Stamens* didynamous, the lower 2 longest, protruding from the corollas or (sub)included, ascending under the upper lip, sticking straight out or divergent; filaments slightly or very unequal in length, $(\frac{1}{50})\frac{1}{10}$ – 1 times as long as the corollas, usually glabrous. *Styles* protruding or (sub)included, c. 1 – $1\frac{1}{2}$ times as long as the corollas. *Nutlets* small, ovoid, brown.

II.3. KEYS TO ALL TAXA EXCEPT HYBRIDS

1. Calyces with 5 (sub)equal teeth .... **Group C**
1. Calyces 2- or 1-lipped
 2. Calyces rather small, 1.3 – 3.5 mm long. Bracts rather small, 1 – 5 mm long, leaf-like in texture and colour, more or less hairy .... **Group B**
 2. Calyces rather large, 4 – 12 mm long. Bracts rather large, 4 – 25 mm long, membranous, usually purple, sometimes yellowish green, more or less glabrous .... **Group A**

Group A

1. Staminal filaments very short, c. 0.5 mm long, all four included. Corollas 15 – 40 mm long, 2-lipped for c. $\frac{1}{7}$; lips nearly at right angles to the tube .... **IV. Longitubus** **20. O. amanum**
1. Staminal filaments 1 – 16 mm long, usually 2 or 4 protruding. Corollas 5 – 17 mm long, 2-lipped for c. $\frac{1}{5}$ or more; tube more or less gradually continuing in the lips
 2. Filaments of upper 2 stamens 1 – 2 mm long, (sub)included; filaments of lower 2 stamens at least 2 × upper ones, (shortly) protruding. Corollas 8 – 16 mm long, 2-lipped for c. $\frac{1}{5}$............................ **III. Brevifilamentum**
  3. Bracts yellowish green. Corollas white or tinged pink
   4. Stems ± hirsute. Leaves usually roundish, with conspicuous veins and few sessile glands .... **19. O. rotundifolium**
   4. Stems ± glabrous. Leaves usually ovate, with inconspicuous veins and many sessile glands .... **14. O. acutidens**
  3. Bracts (partly) purple. Corollas pink
   5. Bracts 5 – 8 × 1 – 3 mm. Spikes 8 – 45 × 4 – 8 mm ... **18. O. leptocladum**
   5. Bracts 5 – 17 × 3 – 13 mm. Spikes 10 – 45 × 9 – 25 mm
    6. Teeth in calyx upper lips obtuse, ± ovate. Filaments of lower 2 stamens up to 9 mm .... **17. O. haussknechtii**
    6. Teeth in calyx upper lips acute, deltoid or triangular. Filaments of lower 2 stamens up to 4 mm
     7. Stems and leaves glabrous. Teeth in calyx upper lips deltoid .... **16. O. brevidens**
     7. Stems and leaves sparingly hirtellous or scabrous. Teeth in calyx upper lips triangular .... **15. O. bargyli**
 2. Filaments of upper 2 stamens 3 – 13 mm long; filaments of lower 2 stamens only slightly longer than upper ones; all 4 stamens (far) protruding. Corollas 5 – 17 mm long, 2-lipped for c. $\frac{1}{3}$
   8. Corollas 8 – 17 mm long, saccate. Calyces 2- or 1-lipped for $\frac{2}{5}$ – $\frac{3}{5}$, with teeth in upper and/or lower lips often reduced or absent. Stamens ascending under

upper lips and far protruding . . . . **I. Amaracus**

9. Flowering stems up to 80 cm long. Full grown stems and leaves ± glabrous
 10. Calyx upper lips (sub)entire. Corollas distinctly saccate . . . . **5. O. saccatum**
 10. Calyx upper lips with 3 deltoid teeth. Corollas (slightly) saccate
  11. Calyx lower lips consisting of 2 narrowly triangular teeth, which are ± as long as upper lips. Corollas saccate . . . . **3. O. cordifolium**
  11. Calyces 1-lipped. Corollas slightly saccate . . . . **6. O. solymicum**
9. Flowering stems up to 40 cm long. Full grown stems and leaves hirsute or (sub)-lanate
 12. Calyx upper lips with 3 and lower lips consisting of 2 acute deltoid or triangular teeth . . . . **1. O. boissieri**
 12. Calyx upper lips (sub)entire; lower lip teeth absent or consisting of 2, c. 0.5 mm long, lobes
  13. Leaves thin, lanate; hairs branched; veins at under side raised. Spikes with short branches, not crowded . . . . **4. O. dictamnus**
  13. Leaves more or less leathery or thin, usually ± hirsute; hairs not branched; veins not raised. Spikes often subsessile and crowded at tops of stems . . . . **2. O. calcaratum**

8. Corollas 5 – 14 mm long, not or barely saccate. Calyces 2-lipped for $\frac{1}{5}-\frac{1}{2}$, with teeth in upper and lower lips usually well developed. Stamens usually ascending under the upper lips and protruding . . . . **II. Anatolicon**
 14. Calyx upper lips (sub)entire
  15. Calyx lower lips consisting of 2 triangular teeth, which are as long as upper lips . . . . **7. O. akhdarense***
  15. Calyx lower lips consisting of 2 ± deltoid, c. 0.5 mm long, teeth . . . . **8. O. cyrenaicum***
 14. Calyx upper lips with 3 deltoid or triangular teeth
  16. Leaves roundish, c. 4 × 3 mm, lanato-pilose. Stems of flowering plants c. 10 cm long . . . . **13. O. vetteri**
  16. Leaves usually ovate, at least c. 10 × 6 mm, glabrous, scabrous or hirtellous. Stems of flowering plants at least 25 cm long
   17. Calyx lower lips consisting of 2 deltoid or triangular teeth, clearly shorter than upper lips. Bracts rather small, c. 7 × 5 mm
    18. Leaves with many sessile glands; lower ones often roundish. Branches up to 3 cm long, not ramified. Bracts often acuminate. Calyx lower lip teeth acute or acuminate . . . . **9. O. hypericifolium**
    18. Leaves with few sessile glands; lower ones ovate. Branches up to 35 cm long, often ramified. Bracts usually obtuse. Calyx lower lip teeth usually obtuse . . . . **12. O. sipyleum**
   17. Calyx lower lip teeth consisting of 2 ± triangular teeth, as long as or slightly shorter than upper lips. Bracts rather large, c. 10 × 7 mm

* For related species from Cyrenaica, **O. pampaninii,** (not included in the key) see p. 127.

19. Leaves more or less heart-shaped, (sub)sessile. Branches up to 4(6) cm long, not ramified .... **11. O. scabrum**
19. Leaves more or less ovate, often petiolate. Branches up to 8 cm long, sometimes ramified .... **10. O. libanoticum**

Group B

1. Tube part of lower lip of calyces nearly absent, so calyces 1-lipped for $\frac{9}{10}$ or more, and bract-like; calyx upper lips (sub)entire.................... **VI. Majorana**
2. Spikes arranged in false corymbs. Leaves often serr(ul)ate...... **26. O. onites**
2. Spikes arranged in panicles. Leaves usually entire
3. Stems and leaves tomentellous (hairs c. 0.3 mm long). Leaves obtuse, veins not raised at under side .... **25. O. majorana**
3. Stems and leaves hirsute, hirsuto-tomentose or tomentose (hairs c. 1 mm long). Leaves usually acute, and usually veins raised at under side **27. O. syriacum**
4. Stems tomentose. Leaves (densely) tomentose, whitish .... **27.a. O. syriacum** var. **syriacum**
4. Stems hirsute. Leaves slightly tomentose or hirsuto-tomentose, greenish
5. Leaves c. 14 × 11 mm, heart-shaped or ovate, shortly petiolate (petioles c. 2 mm long). Corollas c. 4 mm long ...... **27.c. O. syriacum** var. **sinaicum**
5. Leaves c. 25 × 15 mm, (longly) ovate or oval, longly petiolate (petioles c. 5 mm long). Corollas c. 6 mm long..........**27.b. O. syriacum** var. **bevanii**
1. Tube part of lower lip of calyces present, so calyces 1- or 2-lipped for $\frac{1}{5}-\frac{2}{5}$, and tube-shaped; calyx upper lips often with 3 teeth or lobes, and lower lips consisting of 2 teeth or lobes .... **V. Chilocalyx**
6. Leaves 2 – 14 × 2 – 12 mm, (densely) tomentose or tomentellous. Flowers purple or pink
7. Leaves c. 8 × 6 mm, crowded along stems. Calyx upper and lower lips usually with teeth or lobes. Corollas c. 3.5 mm long .... **22. O. micranthum**
7. Leaves c. 5 × 4 mm, not crowded along stems. Calyx upper lips usually entire and lower lips absent. Corollas c. 5 mm long ..... **23. O. microphyllum**
6. Leaves 3 – 23 × 2 – 20 mm, sparsely tomentose, pubescent or pilose. Flowers white
8. Stems and leaves sparsely tomentose or ± pubescent. Bracts 2 – 4 × 1 – 3 mm. Corollas 3 – 6 mm .... **21. O. bilgeri**
8. Stems and leaves ± pilose. Bracts 1 – 3 × 0.5 – 1.5 mm. Corollas 2.5 – 4 mm .... **24. O. minutiflorum**

Group C

1. Calyces (tubular-)campanulate, also when fruit bearing. Spikes rather loose, not clearly demarcated from stems and branches .... **VII. Campanulaticalyx**
2. Corollas c. 3 mm long. Leaves on stems 3 – 7 × 3 – 7 mm . **29. O. isthmicum**

2. Corollas c. 9 mm long. Leaves on stems 3 – 12 × 3 – 8 mm
  3. Stems c. 20 cm long, piloso-tomentose. Calyces c. 4 mm long. Corollas purplish pink .... **30. O. ramonense**
  3. Stems c. 40 cm long, slightly hirsute. Calyces c. 6 mm long. Corollas white .... **28. O. dayi**

1. Calyces tube-shaped, when fruit bearing sometimes turbinate. Spikes loose or dense, but always distinct from stems and branches
  4. Spikes usually longer than 25 mm, loose. Bracts slightly or not imbricate .... **VIII. Elongatispica**
    5. Spikes 4 – 30 × 3 – 5 mm. Bracts 2.5 – 5 × 1 – 3 mm. Leaves pilosellous (hairs c. 0.5 mm long) .... **33. O. grosii**
    5. Spikes 4 – 140 × 3 mm. Bracts 1.8 – 4 × 0.7 – 2 mm. Leaves glabrescent or piloso-tomentose (hairs c. 1 mm long)
      6. Plants glabrescent (only few hairs present), often glaucous. Spikes very lax. Leaves conspicuously glandular punctate, ± leathery ... **31. O. elongatum**
      6. Plants thickly piloso-tomentose, greyish, not glancous. Spikes ± lax. Leaves not conspicuously glandular punctate, not leathery .... **32. O. floribundum**
  4. Spikes usually less than 25 mm long, dense. Bracts imbricate
    7. Corollas 7 – 16 mm long, 2-lipped for c. $\frac{1}{6}$. Stamens (sub)included; filaments c. $\frac{1}{4}$ x corollas .... **X. Prolaticorolla**
      8. Bracts 6 – 11 × 2 – 5 mm, partly purple. Leaves clearly glandular punctate .... **35. O. compactum**
      8. Bracts 3 – 6 × 0.5 – 2.5 mm, partly purple or yellowish green. Leaves glandular punctate or not
        9. Stems and leaves glabrous or slightly scabrous. Brachts partly purple. Sessile glands hardly present at upper side of leaves .... **37. O. laevigatum**
        9. Stems and leaves more or less hirsuto-pilose. Bracts yellowish green. Sessile glands at both sides of leaves present .... **36. O. ehrenbergii**
    7. Corollas 3 – 11 mm long, 2-lipped for c. $\frac{1}{3}$. Stamens (shortly) protruding; filaments c. $\frac{1}{2}$ x corollas .... **IX. Origanum**
34. O. vulgare
      10. Leaves and calyces usually conspicuously glandular punctate. Bracts 1.5 – 6 × 1 – 3 mm
        11. Stems slightly pilosellous or glabrescent. Leaves glandular punctate, ± glaucous, glabrescent or slightly pilosellous. Branches and spikes often slender .... **34.c. O. vulgare** ssp. **gracile**
        11. Stems usually hirsute. Leaves densely glandular punctate, usually not glaucous, usually hirsute or pilosellous. Branches and spikes not slender
          12. Bracts usually shorter than calyces, glabrous or slightly pilosellous along margins, obviously glandular punctate. Inflorescences often very wide ....**34.b. O. vulgare** ssp. **glandulosum**
          12. Bracts usually as long as or somewhat longer than calyces, hirsute or pilosellous, more or less glandular punctate. Inflorescence often com-

pact .... **34.d. O. vulgare** ssp. **hirtum**

10. Leaves and calyces usually inconspicuously glandular punctate. Bracts 2 – 11 × 1 – 7 mm
  13. Bracts usually (partly) purple. Flowers pink .... **34.a. O. vulgare** ssp. **vulgare**
  13. Bracts usually (yellowish) green. Flowers usually white
    14. Bracts 3.5 – 11 × 2 – 7 mm, glabrous or glabrescent, yellowish green. Inflorescences often compact .... **34.e. O. vulgare** ssp. **virens**
    14. Bracts 2 – 8 × 1 – 4 mm, often (densely) pilosellous, usually green. Inflorescences usually not compact .... **34.f. O. vulgare** ssp. **viride**

II.4. DESCRIPTION OF ALL TAXA EXCEPT HYBRIDS

I. **Section Amaracus** (Gleditsch) Bentham

Section *Amaracus* (Gleditsch) Bentham, in de Candolle, Prodr. Syst. Nat. 12: 191 (1848). – Type designated here: *Origanum dictamnus* Linnaeus.
Amaracus section *Euamaracus* (Gleditsch) Briquet, in Engler & Prantl, Nat. Pflanzenfam. 4(3a): 305 (1895).
Subgenus *Amaracus* (Gleditsch) Vogel, Linnaea 15: 76 (1841).

Branches of the first order usually present, those of the second order seldom so. *Leaves* usually leathery. *Spikes* large, usually nodding. *Bracts* imbricate, c. 2 × calyces, membranous, (partly) purple, ± glabrous. *Flowers* usually 2 per verticillaster and subsessile, bisexual, large. *Calyces* tubular 1- or 2-lipped for $\frac{2}{5}$ – $\frac{3}{5}$; teeth in upper and/or lower lips often reduced or absent; throats often pilose. *Corollas* 2-lipped for c. $\frac{1}{3}$ (seldom $\frac{1}{4}$), c. $2\frac{1}{2}$ × calyces, saccate. *Stamens* slightly unequal in length, all 4 ascending under the upper lips and far protruding; filaments ± as long as corollas.

1. **Origanum boissieri** Ietswaart – **Figs. 3, 4** and **5.**

O. boissieri Ietswaart, *nom. nov. O. ciliatum* Boissier et Kotschy, in Boissier, Diagn. Pl. Or. Nov. 2(4): 10 (1859); Boissier, Fl. Or. 4: 548 (1879). *Amaracus ciliatus* (Boissier et Kotschy) Briquet, in Engler & Prantl, Nat. Pflanzenfam. 4(3a): 306 (1895); Bornmüller, Notizbl. Bot. Gart. Berlin 7(63): 26 (1917). – Type: *Kotschy 238*, Turkey, Bulgar Dagh (holo. G, iso. BM, P, W, WU).
Non *O. ciliatum* Willdenow, Car. Linn. Sp. Pl. 3: 133 (1800), (= *Acanthaceae* species). – Type: *Willdenow s.n.*, Guinea (holo. B).

Subshrubs. Roots up to 1 cm in diameter. Young shoots slightly hirsute. Stems usually ascending and rooting at the base, up to 40 cm long, light yellow-brown, slightly villous (hairs c. 1.5 mm long). Branches of the first order always present, in the upper $\frac{2}{5}$ of the stems, up to 8 pairs per stem, 1.2 (0.3 – 4) cm long, not ramified. *Leaves* up to 15 pairs per stem, (sub)sessile, heart-shaped to ovate, tops more or less obtuse or acute, 17 (8 – 30) mm long, 14 (6 – 20) mm wide, thin, light green, more or less glaucous, slightly hirsute (hairs c. 1.5 mm long, mainly at the margins), sessile glands up to 500 per cm^2. *Spikes* subglobose, ovoid or cylindrical, 16 (10 – 25) mm long, 12 (8 – 15) mm wide, more or less nodding. *Bracts* 6 (2 – 10) pairs per spike, (ob)ovate or oval, tops acute to acuminate, 8 (5 – 11) mm long, 5 (3 – 7) mm wide, partly purplish, ciliate at the margins. *Flowers* 2 per verticillaster, (sub)sessile. *Calyces* 2-lipped for c. $\frac{3}{5}$, 5 (4 – 6) mm long, margins and throats pilose; upper lips usually divided, for $\frac{1}{5}$ (but varying), into 3 subequal, ± deltoid, 0.5 (0.1 – 1.2) mm long teeth; lower lips relatively short, $\frac{1}{10}$ – $\frac{1}{5}$ times as long as the upper lips, consisting of 2 (sub)equal deltoid or triangular, 0.8 (0.2 – 1.5) mm long teeth. *Corollas* 2-lipped for c. $\frac{1}{3}$, 13 (10 – 15) mm long, pink, saccate, outside sparsely pilosellous; upper lips divided, for c. $\frac{1}{10}$, into 2, 0.5 (0.1 – 0.8) mm long lobes; lower lips divided for c. $\frac{2}{5}$, into

3 subequal, 1.4 (0.8 – 2.5) mm long lobes. *Staminal filaments* up to 13 and 14 mm long. *Styles* between the filaments protruding under the upper lips, up to 22 mm long.

Geography and ecology. *O. boissieri* has been found in a few places in southern Turkey, in a region formerly named Cilicia, where it has been collected a few times only in the past century. Here it grows or grew from 500 – 2000 m, on rocks and in crevices, usually in shaddy places. It flowers in July and August.
Notes. 1. The ciliate leaves and bracts are remarkable in that they do not occur often in the section *Amaracus*. 2. *O. boissieri* resembles *O. dictamnus* in some characters, but differs from this species in its toothed calyces. 3. Because the existance of an earlier name *O. ciliatum* Willdenow, the species *O. ciliatum* Boissier et Kotschy had to be renamed.

TURKEY. PROV. IÇEL: Taurus Mts., Bulgar Dagh, in valley of Agatsch Kisse, common on shady rocks, 2000 m, 12 Aug. 1853, *Kotschy 238* (type). Taurus Mts., between Gulek-Boghas and Gulek-Maden, in the higher mountain regions, 13 Aug. 1855, *Balansa 542* (G, JE, K, P, W, WU). Dshekeman-Deresi, July 1895, *Siehe 330* (G, JE).

2. **Origanum calcaratum** Jussieu – **Figs. 3, 4** and **5.**

O. calcaratum Jussieu, Gen. Pl.: 115 (1789). – Type: *Jussieu s.n.*, Greece, Amorgos (holo. P).
O. tournefortii Aiton, Hort. Kew. 2: 311 (1789); Sibthorp & Smith, Fl. Graeca 6: 56 (1826); Boissier, Fl. Or. 4: 547 (1879); Halácsy, Consp. Fl. Graec. 2: 553 (1902); Tutin et al., Fl. Eur. 3: 172 (1972). *Amaracus tournefortii* (Aiton) Bentham, Lab. Gen. Sp.: 333 (1834); Hayek, Prodr. Fl. Penins. Balc. 2: 333 (1931); Rechinger, Fl. Aegaea: 531 (1943); Runemark et al., Bot. Not. 113(4): 433 (1960). *Beltokon tourneforti(i)* (Aiton) Rafinesque, Fl. Tell. 3: 86 (1836). – Type: *Sibthorp s.n.*, cultivated from Amorgos (holo. BM).
O. tournefortii Aiton var. *barbatum* Vogel, Linnaea 15: 76 (1841).

Subshrubs. Roots up to 1.5 cm in diameter. Young shoots slightly lanate. Stems erect, up to 35 cm long, light to dark brown, usually hirsute or slightly lanate (hairs c. 1.5 mm long). Branches of the first order nearly always present, in the upper $\frac{2}{5}$ of the stems, up to 6 pairs per stem, often very short, 1 (0.2 – 2.5) cm long, not ramified. *Leaves* up to 35 pairs per stem, (sub)sessile, roundish to ovate or heart-shaped, tops obtuse or acute, 16 (6 – 28) mm long, 14 (5 – 25) mm wide, leathery or thin, hirsute or slightly lanate (hairs c. 1.5 mm long), sometimes glabrous and both sides glaucous, sessile glands often more on the upper sides, up to 600 per cm^2. *Spikes* cylindrical or pyramidal, sometimes subglobose, 25 (10 – 40) mm long, 15 (9 – 17) mm wide, scarcely nodding, often crowded at the top of the stems. *Bracts* 12 (3 – 24) pairs per spike, roundish to oval, tops acute, 10 (5 – 13) mm long, 6 (4 – 10) mm wide, partly slightly purple, glabrous. *Flowers* 2 per verticillaster, subsessile. *Calyces* 1-lipped for c. $\frac{3}{5}$, 6.5 (5 – 8) mm long, throats sparsely pilose or not, otherwise glabrous; upper lips (sub)entire; lower lips absent or consisting of 2 very small, c. 0.5 mm long lobes. *Corollas* 2-lipped for c. $\frac{1}{3}$, 13 (10 – 17) mm long, pink, clearly saccate, outside $\pm$ glabrous; upper lips divided, for c. $\frac{1}{10}$, into 2, 0.5 (0.3 – 1.0) mm long lobes; lower lips divided, for c. $\frac{2}{5}$, into 3 slightly unequal, 2.0 (0.9 – 3.0) mm long lobes. *Staminal filaments* up to 12 and 14 mm long. *Styles* between the filaments protruding under the upper lips, up to 22 mm long. *Chromosome number* 2n = 30.

Geography and ecology. For a long period *O. calcaratum* has been considered as an endemic species of the island of Amorgos, although already in 1855 it has been collected on Nicaria. Recently a much wider area of distribution has been found. In 1938 Barneby and Davis discovered the species on eastern Kriti. Runemark et al. (1960) found it on several other Cyclades: Keros, Anidros, Astipalea, Safora and Sirina. On all these islands it grows from about sea-level up to ca. 700 m, on bare calcareous rocks, which are inaccessible to cattle. It flowers from April to October. Notes. 1. *O. calcaratum* is one of the most variable species in the section *Amaracus*. The following characters are variable: length of stems, ramifications, indumentum of stems and leaves; size, shape, thickness and number of sessile glands of the leaves; size and shape of the bracts; length and shape of corollas. 2. It differs from the related *O. dictamnus*, in the less lanate to glabrous leaves which are also leathery, and in the compact, often pyramidal spikes, which are frequently crowded at the top of the stems. 3. Vogel's var. *barbatum* is not recognized because it is described as differing only in its hairy calyx throats. 4. The name *O. calcaratum* has priority over *O. tournefortii*, because it was published at least a few days earlier. 5. An artificial hybrid is known between *O. calcaratum* and *O. dictamnus* (see p. 141).

GREECE. NICARIA: on rocks, 1 Oct. 1855, *Armenis s.n.* (WU). AMORGOS: cultivated specimens in Kew Gardens, from original collections, 1877, *Sibthorp s.n.* On rocks, April 1872, *Orphanides 1168* (BM, G, P, W, WU). On rocks, 1875, *Orphanides s.n.* (BM, FI, G, WU). On high rocks near the monastery Panagia, 8 Aug. 1881, *Heldreich s.n.* (BM, G, L, W). Mt. Prophet Elias, July 1897, *Leonis s.n.* (G, WU). Near the monastery Panagia Chozoviotissa, 30 June – 6 July 1932, *Rechinger 2314* (BM, K, W). Near the village Langada, in fissures, 30 June – 6 July 1932, *Rechinger 2363* (K, W). Above the monastery Panagia, June 1937, *Guiol 2549* (BM). Near Langada, on rocks, c. 200 m, 6 Oct. 1939, *Davis 944* (E). KRITI: above Sitia, on rocks, c. 700 m, Sept. 1938, *Barneby and Davis s.n.* (E). Sitia, above Rusa Eklesia, alongside road to Krioneri, 500 – 550 m, common in fissures in calcareous rocks, 1 Oct. 1966, *Greuter 7644* (G, W).

3. **Origanum cordifolium** (Montbret et Aucher ex Bentham) Vogel – **Figs. 3, 4** and **5.**

O. cordifolium (Montbret et Aucher ex Bentham) Vogel, Linnaea 15: 76 (1841); Boissier, Fl. Or. 4: 548 (1879). *Amaracus cordifolius* Montbret et Aucher ex Bentham, Ann. Sc. Nat. 2(6): 43 (1836); Bornmüller, Notizbl. Bot. Gart. Berlin 7(63): 26 (1917). – Type: *Aucher s.n.*, Cyprus (holo. W).

Subshrubs. Roots up to 1.5 cm in diameter. Young shoots glabrous. Stems erect, up to 60 cm long, dark or purplish brown, glabrous. Branches of the first order often absent, when present in the upper $\frac{2}{5}$ of the stems, up to 6 pairs per stem, 3 (2 – 6.5) cm long, not ramified. *Leaves* up to 35 pairs per stem, sessile, heart-shaped, tops acute or acuminate, 12 (15 – 20) mm long, 11 (5 – 20) mm wide, entire or remotely serrate, leathery, both sides glaucous, glabrous, sessile glands more at the underside, up to 1500 per cm^2. *Spikes* cylindrical, rather long, 30 (12 – 45) mm (spikes at the end of stems up to 90 mm), 18 (15 – 25) mm wide, nodding. *Bracts* 10 (6 – 12) pairs per spike (on top spikes up to 18 pairs), oval to roundish, tops acuminate, 12 (8 – 17) mm long, 10 (5 – 19) mm wide, partly purple, glabrous. *Flowers* 6 (2 – 20) per verticillaster, with c. 2 mm long pedicels. *Calyces* 2-lipped for c. $\frac{2}{5}$, 6 (5 – 8) mm long, throats pilose, otherwise glabrous; upper lips divided, for c. $\frac{1}{5}$ (but varying), into 3 (sub)equal, ± deltoid, 0.5 (0.2 – 1.0) mm long teeth; lower lips approximately as long as the upper lips, consisting of 2 (sub)equal (narrowly) triangular, 2.0 (1.5 – 3.0) mm long teeth. *Corollas* 2-lipped for c. $\frac{1}{3}$, 12 (10 – 13) mm long, pink, saccate, outside pilosellous; upper lips divided, for c. $\frac{1}{10}$, into 2, 0.3 (0.2 – 0.5) mm long lobes; lower lips divided, for c. $\frac{2}{5}$, into 3 subequal, 1.4 (0.8 – 1.8) mm long lobes. *Staminal filaments* up to 10 and 11 mm long. *Styles* between the filaments protruding under the upper lips, up to 16 mm long.

Geography and ecology. *O. cordifolium* occurs on Cyprus, in some places in the Kikko Mts. (part of the Troodos massif). There is one report (Aucher Eloy) from Syria, without any further indication of the site. On Cyprus the species is found from 500 – 1500 m, especially on rocks, in crevices and along mountain streams. Sometimes it grows in the shadow of pine trees. It flowers from April to August.

Notes. 1. *O. cordifolium* has some characters that are uncommon in the section *Amaracus*: several flowers per verticillaster, long spikes, and calyx lower lips with 2 (narrowly) triangular teeth, which are approximately as long as the upper lips. 2. *O. cordifolium* resembles in general habit and in the absence of an indumentum *O. saccatum* and *O. solymicum*, but it differs from these species in the characters just mentioned.

CYPRUS: 1833, *Aucher s.n.* (type). Ticco, 1150 – 1450 m, 25 – 30 June 1913, *Haradjian 954* (G, K). Roudhkias Valley, in crevices, 850 m, 10 Aug. 1938, *Chapman 348* (K). Roudhkias Valley between Pano Panorgia and Kykko, cracks of igneous rocks above perpetual stream in shady ravine, 500 – 700 m, 8 May 1941, *Davis 3388* (K). Kykko Mt., Audkias basin, 700 m, rocky sides of a mountain stream, 9 July 1948, *Kennedy 1633* (B, F1, G, K, L). Roudhkias Valley, Papheto forest, abundant on dry rocky slopes in pine forest or on rock faces, 8 Aug. 1954, *Merton 1944* (K).

SYRIA: 1837, *Aucher 1656* (FI, G, P, W).

4. **Origanum dictamnus** Linnaeus – **Figs. 3, 4** and **5.**

O. dictamnus Linnaeus, Sp. Pl.: 589 (1753); Chaubard & Bory, Nouv. Fl. Péloponn. Cycl.: 38 (1838); Boissier, Fl. Or. 4: 547 (1879); Halácsy, Consp. Fl. Graec. 2: 552 (1902); Tutin et al., Fl. Eur. 3: 172 (1972). *Majorana dictamnus* (Linnaeus) Kosteletzky, Allg. Med.-Pharm. Fl. 3: 770 (1834). *Amaracus dictamnus* (Linnaeus) Bentham, Lab. Gen. Sp.: 333 (1834); Gams, in Hegi, Ill. Fl. Mittel-Eur. 5: 2333 (1927); Hayek, Prodr. Fl. Penins. Balc. 2: 333 (1931); Rechinger, Fl. Aegaea: 531 (1943); Lawrence Baileya 2: 30 (1954); Wolf, Baileya 2: 57 (1954). – Type: *Linnaeus 743.2* (holo. LINN).

Dictamnus creticus Hill, Br. Herb.: 381 (1756).

O. saxatile Salisbury, Prodr. Stirp.: 85 (1796).

Amaracus tomentosus Moench, Meth., Suppl.: 137 (1802).

Majorana tomentosa Stokes, Bot. Mat. Med. 3: 348 (1812).

O. pseudodictamnus Sieber, Flora 1: 273 (1818).

O. dictamnifolium Saint-Lager, Ann. Soc. Bot. Lyon 7: 131 (1880).

Subshrubs. Roots up to 1 cm in diameter. Young shoots lanate. Stems ascending and rooting at the bases, up to 35 cm long, yellow or purplish brown, lanate (hairs c. 2 mm long, branched). Branches of the first order absent or present, in the upper $\frac{1}{2}$ of the stems, up to 5 pairs per stem, 1.5 (0.5 – 4) cm long, not ramified. *Leaves* up to 15 pairs per stem, the lower ones petiolate (petioles up to 15 mm long), roundish to oval or ovate, tops obtuse or acute, 15 (4 – 30) mm long, 15 (4 – 30) mm wide, thin, whitish (green), both sides lanate (hairs c. 2 mm long, branched), sessile glands inconspicuous, up to 600 per cm^2; margins revolute; veins raised at the underside. *Spikes* subglobose to cylindrical, 16 (7 – 40) mm long, 14 (10 – 17) mm wide, nodding. *Bracts* 8 (3 – 16) pairs per spike, roundish to (ob)ovate, tops obtuse or acute, 9 (6 – 12) mm long, 7 (4 – 10) mm wide, more or less purple, glabrous or sparsely ciliate. *Flowers* 2 per verticillaster, subsessile. *Calyces* 1-lipped for c. $\frac{3}{5}$, 5 (4 – 7) mm long, throats sparsely pilose or not, for the rest glabrous; upper lips (sub)-entire; lower lips absent or consisting of 2 very small (c. 0.5 mm long) lobes. *Corollas* 2-lipped for c. $\frac{1}{3}$, 11 (8 – 15) mm long, pink, more or less saccate, outside sparsely pilosellous; upper lips divided, for c. $\frac{1}{10}$, into 2, 0.3 (0.2 – 0.5) mm long lobes; lower lips divided, for c. $\frac{2}{5}$, into 3 subequal, 1.5 (1.0 – 2.8) mm long lobes. *Staminal filaments* up to 12 and 14 mm long. *Styles* between the filaments protruding under the upper lips, up to 18 mm long.

Geography and ecology. *O. dictamnus* occurs on several places on Kriti, while Rechinger mentioned it for Poros too. In 1969 Fokkinga, who studied some *Origanum* species in Greece, could not rediscover the species there, in spite of intensive search (pers. comm.). In 1886 Forsyth-Major collected specimens of an *Origanum* species in Mykali (western Turkey, opposite Samos) that later on have been named *O. dictamnus*. The scanty material from this collection that could be studied lacks the characteristic branched hairs. Flowers could not be studied. Perhaps a new species is concerned here. On Kriti *O. dictamnus* can be found from 300 – 1500 m, on calcareous rocks, debris and in cracks, often on shady places. It flowers from June to October.

Notes. 1. *O. dictamnus* is unique in the genus *Origanum* for its branched hairs. 2. *O. dictamnus* is related to *O. calcaratum*, from which it differs in its lanate and thinner leaves, its less compact inflorescences and usually subglobose spikes. 3. Chaubard & Bory (1838) mentioned *O. dictamnus*, but in fact described the morphological characters as well as the type locality of *O. calcaratum*. For this reason *O. dictamnus* Chaubard et Bary, sometimes is found cited as a synonym of *O. calcaratum*. 4. Three hybrids, all of garden origin, are known of which *O. dictamnus* as one of the parents. In recent times artificial hybrids have been made with *O. amanum* and *O. calcaratum* (pp. 140 and 141). The third hybrid, *O.* × *hybridinum*, with probably as other parent *O. sipyleum*, was already mentioned by Miller (p. 136).

GREECE. KRITI: Mt. Sfakia, on shady rocks and in fissures, July 1846, *Heldreich s.n.* (BM, P, W). Temenos, in fissures, 30 Aug. 1873, *Heldreich s.n.* (BM, FI, JE). Nipros, 20 July 1882, *Spreizenhofer s.n.*

(W, WU). Akroteri, on rocks, 20, 23 en 29 June 1883, *Reverchon s.n.* (BM, G, W). Along the road from Kavosi to Asomatos, on steep rocks, 12 Aug. 1893, *Baldacci 190* (BM, G). Hagios Vasilis, Spili, on rock-face, 27 June 1904, *Dörfler 663* (WU). Monophatsi, Mt. Kophina, 6 July 1904, *Dörfler 880* (WU). Lasithi, Mt. Aphendi Khristos, rock-face, 27 July 1904, *Dörfler 1045* (WU). Hierapetra, Mt. Aphendi Kavusi, on bare rocks, 800 m, 1 Aug. 1904, *Dörfler 1089* (B, JE, G, W, WU). Idem, *Dörfler 4755* (E, FI, G, JE, P, W, WU). Mt. Lakous, June 1932, *Guiol 2120* (BM). Asilositis, July 1934, *Regel s.n.* (G). Lasithi, July 1934, *Regel s.n.* (G). Chania, peninsula Akroteri, in fissures in calcareous rocks, 25 May 1942, *Rechinger 13315* (W). Selinos, near Palaeochora, in fissures in calcareous rocks, 1 June 1941, *Rechinger 13531* (W). Sfakia, pass of Imvros, above the road, c. 650 m, in fissures in calcareous rocks, 18 Sept. 1966, *Greuter 7536* (E, G). Ierapetra-Sitia, Mt. Aphendi Kavusi, n. facing slopes, c. 1150 m, crevices in calcareous rocks, 29 Sept. 1966, *Greuter 7627* (G). Sfakia, Asfendos, fissures in steep calcareous rocks, c. 600 m, 8 Oct. 1966, *Greuter 7688* (G). Ierapetra, between Refkos and Murnies, c. 500 m, on and between boulders of calcareous rock, 27 Oct. 1966, *Greuter 7795* (G).

5. **Origanum saccatum** Davis – **Figs. 5** and **6.**

O. saccatum Davis, Kew Bull. 1949: 409 (1949). – Types: *Davis 14276*, Turkey, Antalya, Kozlu Dere (holo. K, iso. E); *Davis 14397*, Turkey, Antalya, Kargi Çay (para. E, G, K).

Subshrubs. Roots up to 1 cm in diameter. Young shoots hirsute. Stems erect, up to 80 cm long, light or yellow brown, at the bases somewhat hirsute (hairs c. 1 mm long), otherwise glabrous. Branches of the first order always present, in the upper $\frac{1}{2}$ of the stems, up to 12 pairs per stem, 4 (0.5 – 6.5) cm long, not ramified or sometimes a few branches of the second order present. *Leaves* up to 30 pairs per stem, (sub)sessile, heartshaped to ovate, tops acute or obtuse, 20 (6 – 40) mm long, 13 (4 – 24) mm wide, leathery, both sides glaucous, $\pm$ glabrous, sessile glands up to 600 per cm^2. *Spikes* oblong-ovoid to cylindrical, 20 (10 – 35) mm long, 7 (6 – 11) mm wide, nodding. *Bracts* 8 (4 – 14) pairs per spike, ovate, tops acuminate or acute, 9 (6 – 11) mm long, 3 (2 – 6) mm wide, partly bright purple, glabrous. *Flowers* 2 per verticillaster, (sub)sessile. *Calyces* 1-lipped for c. $\frac{2}{5}$, 5 (4 – 6) mm long, throats pilose, otherwise glabrous; upper lips (sub)entire; lower lips absent or consisting of 2 very small lobes. *Corollas* 2-lipped for c. $\frac{1}{3}$, 12 (11 – 13) mm long, pink, clearly saccate, outside sparsely pilosellous; upper lips divided, for c. $\frac{1}{10}$, into 2, 0.5 (0.3 – 1.2) mm long lobes; lower lips divided, for c. $\frac{2}{5}$, into 3 slightly unequal, 1.6 (1.1 – 2.8) mm long lobes. *Staminal filaments* up to 11 and 12 mm long. *Styles* between the filaments protruding under the upper lips, up to 20 mm long.

Geography and ecology. *O. saccatum* occurs in a few places in southern Turkey, in the provinces of Antalya and Isparta. It grows on calcareous rocks and slopes, sometimes under pine trees, at c. 1000 m. It flowers in July and August.
Note. *O. saccatum* is rather closely related to *O. solymicum* and somewhat to *O. sipyleum*. It differs from both species in its non-toothed calyces and its clearly saccate corollas.

TURKEY. PROV. ANTALYA: Kargi Çay near Kozlu Dere (n.e. of Alanya), 1000 m, 26 Aug. 1947, *Davis 14276* (type). Kargi Çay between Durbanas and Derince Dere, rocky limestone slopes in pine woods, c. 1000 m, 24 Aug. 1947, *Davis 14397* (paratype). Between Kizil Alan and Durbanas, 24 Aug. 1947, *Davis*

14439 (E, K). Kargi Çay near Kozlu Dere (n.e. of Alanya), 1000 m, 26 Aug. 1947, *Davis 14276* (E). PROV. ISPARTA: between Daribükü and Selköse, 30 July 1949, *Davis 15857*, non flowering specimen (E).

6. **Origanum solymicum** Davis – **Figs. 3, 4** and **5.**

O. solymicum Davis, Kew Bull. 1949: 410 (1949). – Types: *Davis 14078*, Turkey, Antalya, Tahtali Dağ (holo. K, iso. E); *Davis 14099*, Turkey, Antalya, Kesme Boğaz (para. E).

Subshrubs. Young shoots hirsute. Stems erect, up to 80 cm long, light to dark brown, at the bases somewhat hirsute (hairs c. 1 mm long), otherwise glabrous. Branches of the first order absent or present, in the upper $\frac{2}{5}$ of the stems, up to 8 pairs per stem, 3 (0.5 – 7) cm long, not ramified. *Leaves* up to 25 pairs per stem, (sub)sessile, heart-shaped to ovate, tops acute or obtuse, 18 (5 – 22) mm long, 14 (3 – 18) mm wide, leathery, both sides glaucous, glabrous, sessile glands up to 400 per cm^2. *Spikes* oblong-ovoid to cylindrical, 25 (10 – 50) mm long (those at the end of branches often much longer), 11 (9 – 15) mm wide, nodding. *Bracts* 8 (5 – 14) pairs per spike, ovate, tops acuminate or acute, 10 (8 – 13) mm long, 6 (4 – 8) mm wide, partly slightly purple, glabrous. *Flowers* 2 per verticillaster, subsessile. *Calyces* 1-lipped for c. $\frac{2}{5}$, 6 (5 – 6) mm long, throats pilose, otherwise glabrous; upper lips divided, for c. $\frac{1}{5}$, into 3 subequal, deltoid, 0.8 (0.5 – 1.5) mm long teeth; lower lips absent. *Corollas* 2-lipped for c. $\frac{1}{4}$, 13 (11 – 15) mm long, pink, more or less saccate, outside ± glabrous; upper lips divided, for c. $\frac{1}{5}$, into 2, 0.8 (0.5 – 1.5) mm long lobes; lower lips divided, for c. $\frac{2}{5}$, into 3 subequal, 1.4 (1.0 – 2.4) mm long lobes. *Staminal filaments* up to 12 and 14 mm long. *Styles* between the filaments protruding under the upper lips, up to 20 mm long.

Geography and ecology. On the only site where *O. solymicum* has been found until now it grows on calcareous rocks and slopes, sometimes in half shade of *Pinus brutia*.

Note. From related species, like *O. saccatum* and *O. cordifolium*, *O. solymicum* differs in its rather long tubed corolla, which is only slightly saccate, and in its toothed calyx upper lips combined with missing lower lips.

TURKEY. PROV. ANTALYA: mountains w. of the gulf of Antalia, near Schizali, on rocky, bare places, Aug. 1863, *unknown collector s.n.* (FI). Tahtali Dağ, between Kuzdere and Kesme Boğaz, on calcareous rocks in Pinetum brutiae, 15 Aug. 1947, *Davis 14078* (type). Kesme Boğaz near Kemer, 60 m, in river bed, 17 Aug. 1947, *Davis 14099* (paratype).

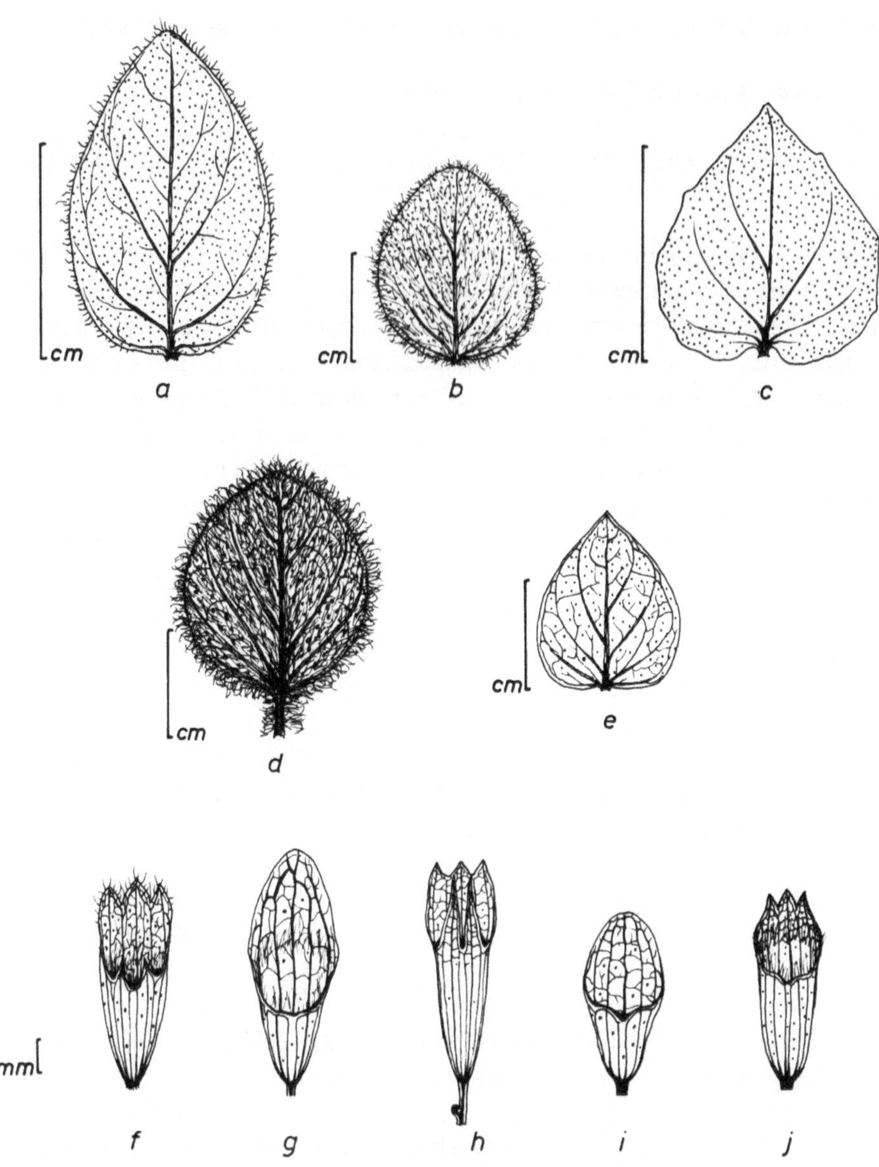

Figure 3. Leaves and calyces of the species in the section *Amaracus*, except *O. saccatum* (for which see figure 6): a. and f. *O. boissieri;* b. and g. *O. calcaratum;* c. and h. *O. cordifolium;* d. and i. *O. dictamnus;* e. and j. *O. solymicum.*

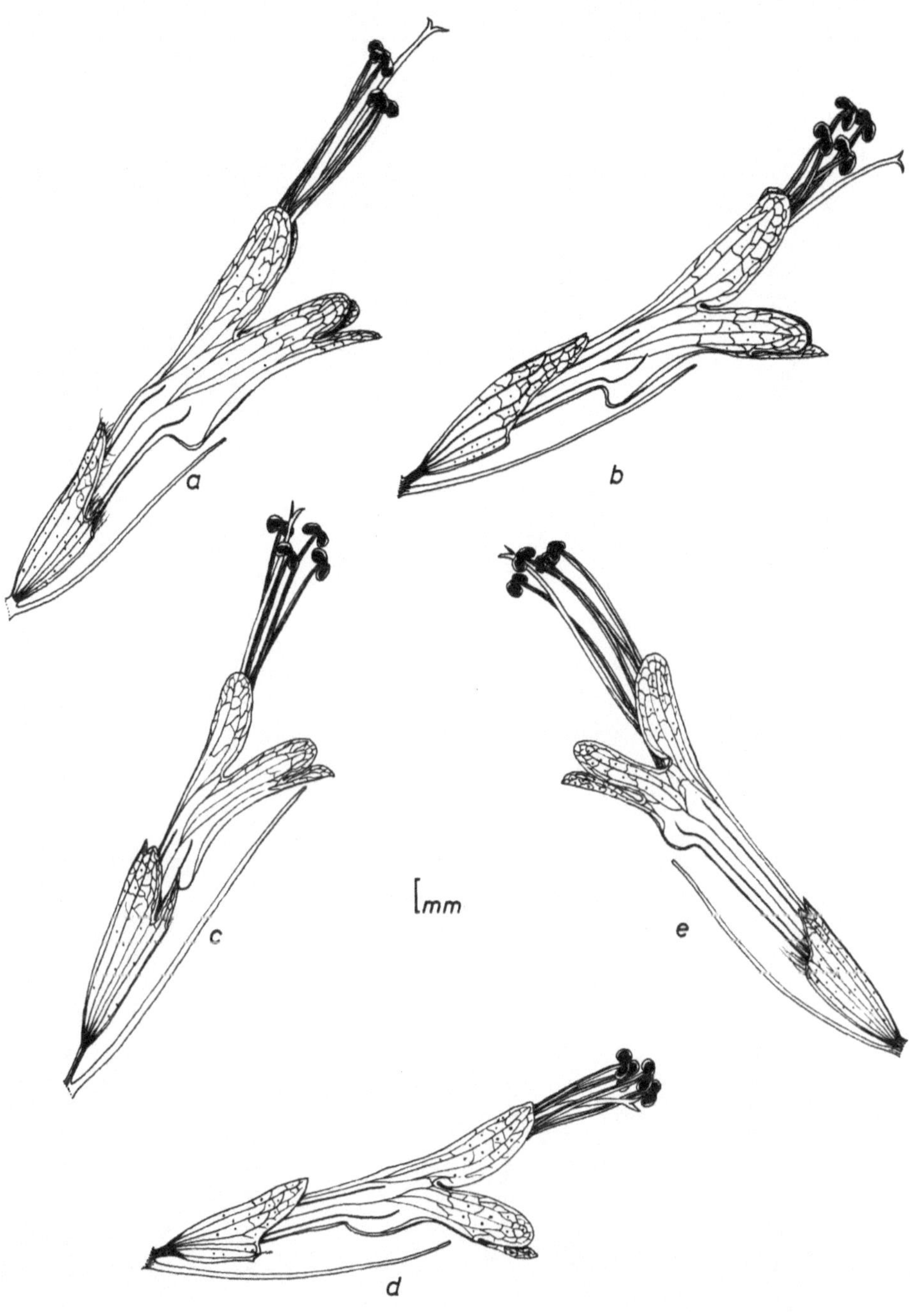

Figure 4. Flowers with bracts in side view of the species in the section *Amaracus*, except *O. saccatum* (for which see figure 6): a. *O. boissieri;* b. *O. calcaratum;* c. *O. cordifolium;* d. *O. dictamnus;* e. *O. solymicum*.

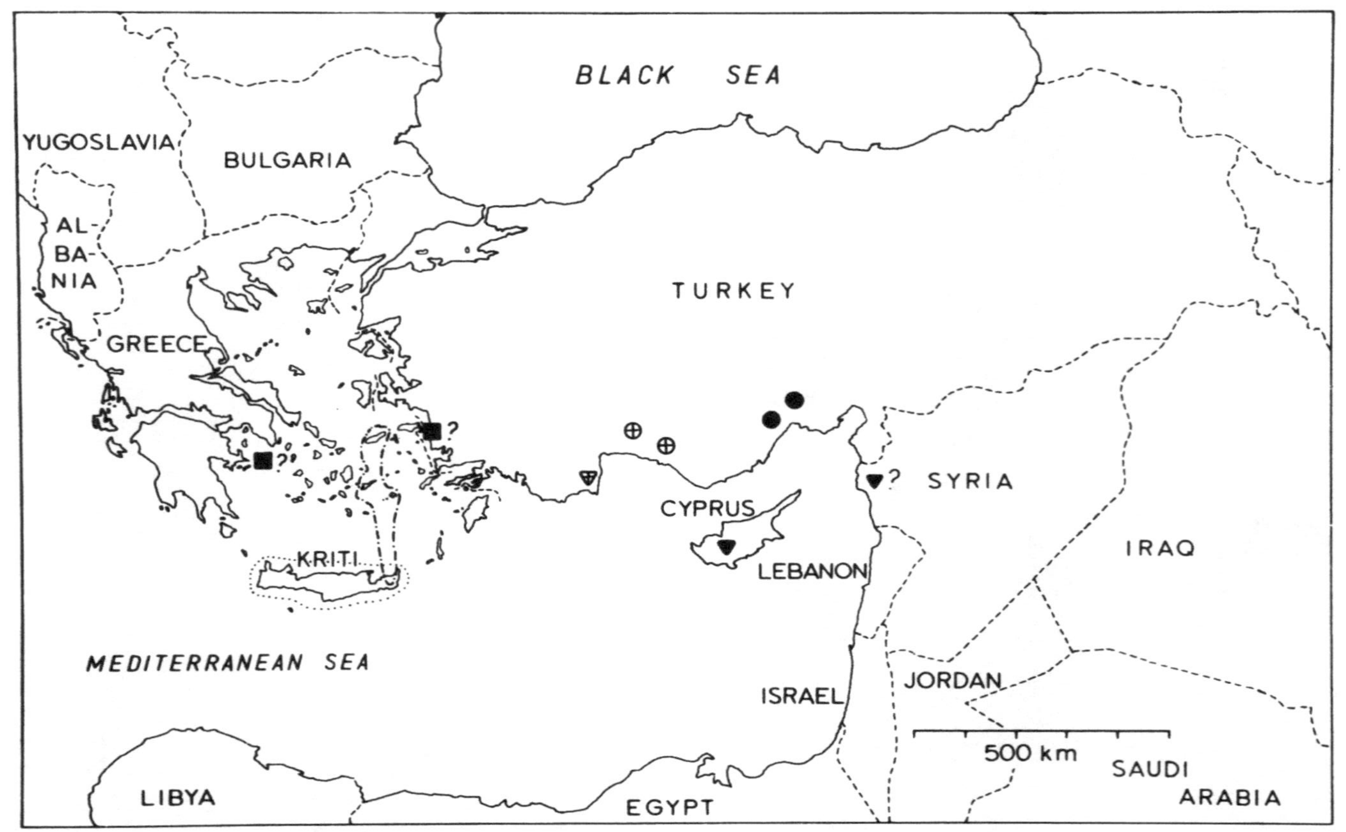

Figure 5. Distribution of the species in the section *Amaracus*: ● *O. boissieri*; —·—·— *O. calcaratum*; ▼ *O. cordifolium*; .... and ■ *O. dictamnus*; ⊕ *O. saccatum*; ∀ *O. solymicum*; ? based on one doubtful record.

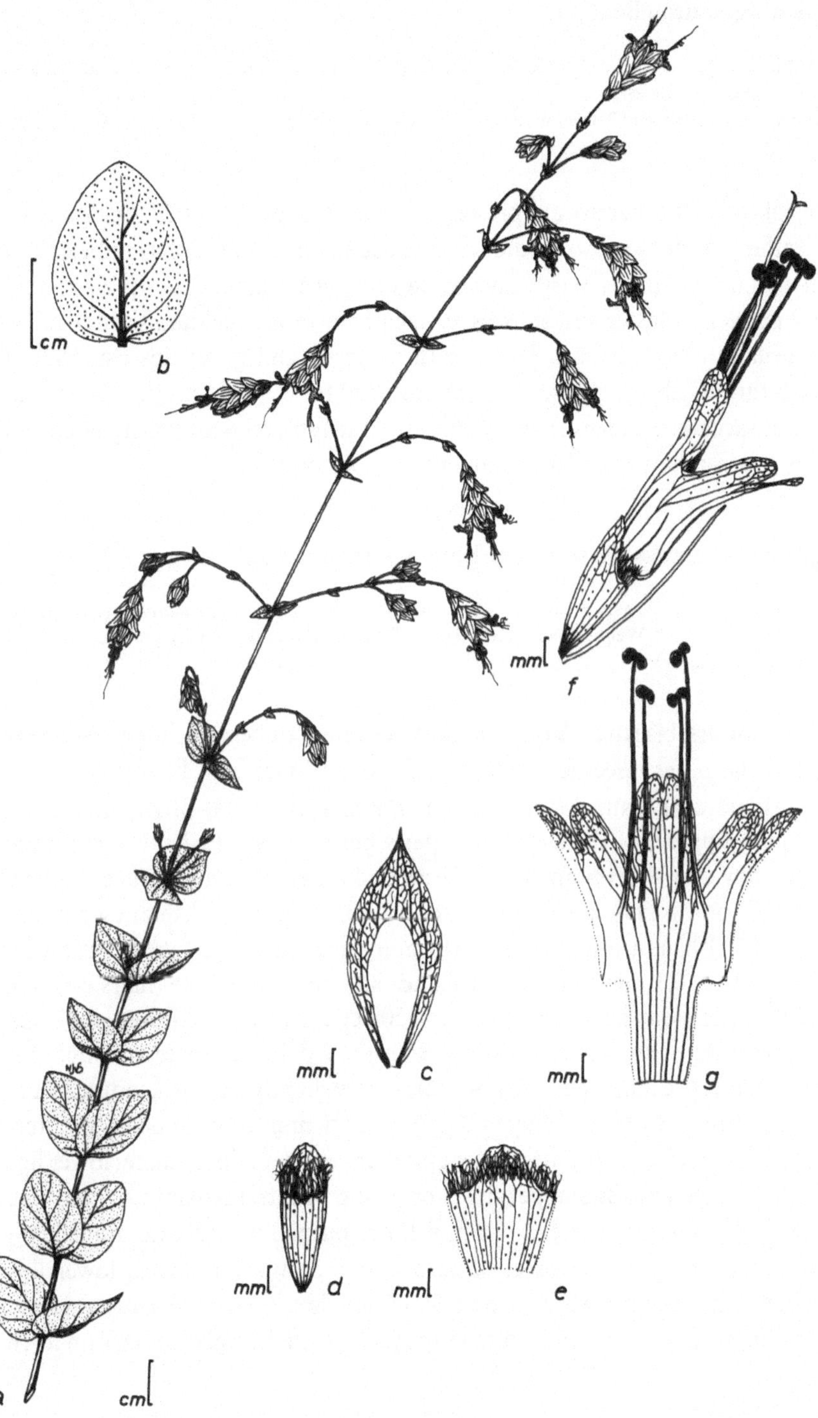

Figure 6. O. saccatum: a. habit; b. leaf; c. bract with calyx in upper lip view; d. calyx in lower lip view; e. calyx cut through the lower lip; f. flower with bract in side view; g. corolla cut through the lower lip.

II. **Section Anatolicon** Bentham

Section *Anatolicon* Bentham, in de Candolle, Prodr. Syst. Nat. 12: 192 (1848). – Type designated here: *Origanum libanoticum* Boissier.
Amaracus section *Anatolicon* (Bentham) Briquet, in Engler & Prantl, Nat. Pflanzenfam. 4(3a): 305 (1895).

Branches of the first order always present, those of the second order sometimes so. *Leaves* leathery or not. *Spikes* medium sized, usually nodding. *Bracts* imbricate, c. $1\frac{1}{2}$ × calyces, membranous, partly and/or slightly purple, glabrous or pilose(llous). *Flowers* 2 per verticillaster and subsessile, usually bisexual, medium sized. *Calyces* tubular, usually 2-lipped for c. $\frac{1}{5}$ – $\frac{1}{2}$; teeth in upper and lower lips usually well developed; throats always pilose. *Corollas* 2-lipped for c. $\frac{1}{3}$ (seldom $\frac{1}{4}$), 2 × calyces, (nearly) not saccate. *Stamens* slightly unequal in length, all 4 ascending under the upper lips or straight, protruding; filaments c. $\frac{3}{5}$ × corollas.

7. **Origanum akhdarense** Ietswaart et Boulos – **Figs. 7, 8** and **9.**

O. akhdarense Ietswaart et Boulos, Acta Bot. Neerl. 24: 285 (1975). *Amaracus akhdarensis* (Ietswaart et Boulos) Brullo et Furnari, Webbia 34(1): 440 (1979). – Type: *Boulos 4469*, Libya, Cyrenaica, Wadi El-Kouf (holo. CAI, iso. AVU).

Subshrubs, flowers bisexual. Young shoots minutely pubescent. Stems sparsely branched at the bases, erect or ascending, c. 30 cm long, light brown, minutely pubescent (hairs c. 0.1 mm long, partly glandular). Branches of the first order present, in the upper $\frac{2}{5}$ of the stems, c. 7 pairs per stem, 0.6 (0.2 – 1.4) cm long; sometimes a few very short branches of the second order present. *Leaves* c. 14 pairs per stem, more or less petiolate (petioles up to 3 mm long), roundish or ovate, tops obtuse, 16 (8 – 19) mm long, 13 (6 – 16) mm wide, thin, light green, minutely pubescent (hairs c. 0.1 mm long, partly glandular), margins ciliate (hairs c. 1 mm long), sessile glands scarsely apparent, up to 150 per cm^2. *Spikes* ellipsoid, c. 12 mm long, c. 6 mm wide, more or less nodding. *Bracts* c. 7 pairs per spike, oblong, tops more or less acute, c. 6 mm long, c. 2 mm wide, partly slightly purplish, indumentum ± as in the leaves. *Calyces* 2-lipped for c. $\frac{1}{5}$, c. 5 mm long, outside minutely pubescent, margins ciliate, throats pilose; upper lips entire or denticulate; lower lips approximately as long as the upper lips, consisting of 2 triangular, c. 1 mm long teeth. *Corollas* 2-lipped for c. $\frac{1}{4}$, c. 11 mm long, purple, not saccate, outside ± glabrous; upper lips divided, for c. $\frac{1}{10}$, into 2, c. 0.3 mm long lobes; lower lips divided, for c. $\frac{3}{5}$, into 3, somewhat unequal, c. 1.7 mm long lobes. *Stamens* straightly protruding; filaments up to 6 and 8 mm long. *Styles* straightly protruding, up to 16 mm long.

Geography and ecology. *O. akhdarense* seems to be confined to the Wadi El-Kouf which borders the distribution area of *O. cyrenaicum*, according to Brullo & Furnari (1979). Here it occurs frequently on shady rock faces together with several other relic endemic species. It has been found flowering in September and October.

Notes. 1. *O. akhdarense* differs from *O. cyrenaicum* in its minutely pubescent, thin leaves, smaller, ciliate bracts, clearly bidentate calyx lower lips, that are approximately as long as the upper lips, and its purple corollas. From the related *O. pampaninii* (which is treated on p. 127) it differs in its less hairy habit, its somewhat smaller bracts and flowers and in its purple corollas with 4, long, protruding stamens. 2. According to the authors just mentioned *O. akhdarense* is exclusively autumn flowering, and not producing a strong smell (like most other *Origanum* species).

LIBYA. CYRENAICA: Gebel Akhdar, Wadi El-Kouf, 27 Oct. 1970, *Boulos 4469* (type). Wadi El-Kouf, 14 Sept. 1974, *Brullo & Furnari s.n.* (CAT).

8. **Origanum cyrenaicum** Béguinot et Vaccari – **Figs. 7, 8** and **9.**

O. cyrenaicum Béguinot et Vaccari. Ann. Bot. Roma 12: 117 (1913); Pampanini, Prodr. Fl. Cirenaica: 400 (1931). *Amaracus cyrenaicus* (Béguinot et Vaccari) Rechinger, Vegetatio 2: 254 (1950), (nomen invalidum). *Amaracus cyrenaicus* (Béguinot et Vaccari) Rechinger ex Brullo et Funari, Webbia 34(1): 442 (1979). – Type: *Béguinot & Vaccari 180*, Libya, Cyrenaica, Wadi Derna (holo. PAD, iso. FI, E).

Subshrubs, flowers bisexual. Roots up to 1 cm in diameter. Young shoots hirsute to scabrous. Stems usually ascending and rooting at the bases, up to 50 cm long, yellowish or purplish brown, scabrous to hirsute (hairs 0.1 – 2.0 mm long) or glabrous. Branches of the first order present, in the upper $\frac{2}{5}$ of the stems, up to 8 pairs per stem, relatively short, 1 (0.4 – 4) cm long; sometimes short branches of the second order present. *Leaves* up to 25 pairs per stem, the lower ones shortly petiolate (petioles up to 5 mm long), hearth-shaped to ovate, tops obtuse or acute, 11 (3 – 18) mm long, 7 (2 – 14) mm wide, ± leathery, light green or purplish, scabrous to hirsute (hairs 0.1 – 1.5 mm long), or glabrous, sessile glands inconspicuous, up to 400 per cm^2. *Spikes* ± ellipsoid, 12 (6 – 20) mm long, 7 (6 – 14) mm wide, more or less nodding. *Bracts* 8 (4 – 13) pairs per spike, ovate to oval, tops ± acute, 8 (4 – 13) mm long, 5 (2 – 20) mm wide, partly slightly purple, sparsely pilose or glabrous. *Calyces* 2-lipped for c. $\frac{1}{2}$, 5 (4 – 6) mm long, throats pilose, otherwise somewhat pilosellous; upper lips (sub)entire; lower lips relatively very short, consisting of 2 equal, c. 0.5 mm long, ± deltoid teeth. *Corollas* 2-lipped for c. $\frac{1}{4}$, 10 (8 – 13) mm long, white, not saccate, outside very sparsely pilosellous; upper lips divided, for c. $\frac{1}{10}$, into 2, 0.3 (0.1 – 0.6) mm long lobes; lower lips divided, for c. $\frac{3}{5}$, into 3 subequal, 1.5 (1.0 – 2.1) mm long lobes. *Stamens* straightly protruding; filaments up to 5 and 7 mm long. *Styles* straightly protruding, up to 16 mm long.

Geography and ecology. *O. cyrenaicum* is an endemic species of Cyrenaica. It is found on hills along several "wadi's", on calcareous rocks, from c. 200 – 600 m. Here it could not be rediscovered recently by Brullo & Furnari (1979). It flowers from May to September.

Notes. 1. *O. cyrenaicum* possesses, like the other two Cyrenaic species (*O. akhdarense* and *O. pampaninii*), some characters not often found in the section *Anatolicon* (and a few of them even not in the genus *Origanum*). For example a very clear chamaephytic habit, frequent occurrence of non-flowering axillary shoots in the lower parts of the stems, short spike-bearing branches that persist after the bracts and flowers have fallen off, small bracts that are only slightly purple coloured, and (sub)entire calyx upper lips. 2. From the two species just mentioned *O. cyrenaicum* differs in its calyces which are 2-lipped to c. $\frac{1}{2}$ and which possess very short lower lips.

LIBYA. CYRENAICA: Derna, on calcareous hills along right side of Wadi Garabba, near mouth of the Wadi Derna, 20 June 1912, *Béguinot & Vaccari 180* (type). Wadi Nardana, June 1919, *Maugini s.n.* (FI, K). Wadi Sneidi, June 1919, *Maugini s.n.* (FI). Wadi Belgadir, 16 Apr. 1934, *Pampanini & Pichi-Sermolli 6704* (FI). Near Cirene, 29 Apr. 1934, *Pampanini & Pichi-Sermolli 6705* (FI). El-Beda, Wadi Urdama, 2 May 1934, *Pampanini & Pichi-Sermolli 6706* (FI). El-Beda, Wadi Scisu, 7 May 1934, *Pampanini & Pichi-Sermolli 6707* (FI). El-Beda, Wadi Msuria, 10 May 1934, *Pampanini & Pichi-Sermolli 6708* (FI). Cirene, Wadi Bu Nabeh, 15 May 1934, *Pampanini & Pichi-Sermolli 6709* (FI). Wadi Wardama, near Littoria, on limestone rocks of deep gorge, c. 600 m, 6 Apr. 1939, *Sandwith 2423* (K).

9. **Origanum hypericifolium** Schwarz et Davis – **Figs. 7, 8** and **9.**

O. hypericifolium Schwarz et Davis, Kew Bull. 1949: 407 (1949). – Types: *Davis 13401*, Turkey, Denizli, Boz Dağ (holo. K, iso. E); *Davis 13599*, Turkey, Muğla, Sandras Dağ (para. E, K).

Subshrubs, flowers bisexual. Roots c. 0.5 cm in diameter. Young shoots hirtellous. Stems ascending and rooting at the base or erect, up to 50 cm long, yellowish or purplish brown, hirtellous overall or only at the bases (hairs c. 0.4 mm long). Branches of the first order present, in the upper $\frac{2}{5}$ of the stems, up to 7 pairs per stem, 1.2 (0.5 – 3) cm long, not ramified. *Leaves* up to 28 pairs per stem, subsessile, (longly) ovate, the lower ones often roundish, tops acute, 15 (4 – 20) mm long, 7 (2 – 12) mm wide, ± leathery, glaucous, more or less hirtellous to scabrous (hairs c. 0.2 mm long), relatively many sessile glands, up to 1200 per cm^2. *Spikes* subglobose or ellipsoid, 11 (8 – 20) mm long, 8 (5 – 12) mm wide, usually nodding. *Bracts* 7 (4 – 9) pairs per spike, roundish to oval, tops ± acuminate, 8 (6 – 10) mm long, 5 (4 – 7) mm wide, partly purple, glabrous. *Calyces* 2-lipped for c. $\frac{2}{5}$, 5 (4 – 6) mm long, throats pilose, for the rest glabrous; upper lips divided, for c. $\frac{2}{5}$, into 3 (sub)equal ± triangular 0.7 (0.5 – 1.0) mm long teeth; lower lips rather short, consisting of 2 (sub)equal, deltoid or triangular, 0.5 (0.2 – 0.8) mm long teeth. *Corollas* 2-lipped for c. $\frac{1}{3}$, 9 (8 – 11) mm long, pink, (nearly) not saccate, outside sparsely pilosellous; upper lips divided, for c. $\frac{1}{5}$, into 2, 0.8 (0.6 – 1.0) mm long lobes; lower lips divided, for c. $\frac{3}{5}$, into 3 subequal, 1.8 (1.4 – 2.3) mm long lobes. *Stamens* ascending under the upper lips; filaments up to 7 and 8 mm long. *Styles* between the filaments protruding under the upper lips, up to 13 mm long.

Geography and ecology. *O. hypericifolium* is known only from some mountains in southwestern Turkey. Here it is found from 1500 – 2000 m, on rocky places, partly in black pine forest. It has been found flowering in July and August.

N o t e. *O. hypericifolium* is rather closely related to *O. sipyleum*, from which it differs in its much more glandular punctate leaves, its acuminate bracts and its more acute calyx teeth. From *O. scabrum* it differs in its more narrow leaves and bracts, and from *O. libanoticum* in its shorter calyx lower lips.

TURKEY. PROV. DENIZLI: Honaz Dağ near Honaz, on slopes, 14 Aug. 1932, *Regel s.n.* (B, G). Boz Dağ above Geyran Yaylâ, 1500 – 2000 m, 16 July 1947, *Davis 13401* (type). PROV. MUGLA: Pirnas Dağ, near Pirnas, in open Pinetum, on limestone, c. 1300 m, 5 Aug. 1938, *Schwarz 443* (JE). Sandras Dağ above Ağla, 1500 m, 25 July 1947, *Davis 13599* (paratype).

10. **Origanum libanoticum** Boissier – **Figs. 9** and **10.**

O. libanoticum Boissier, Diagn. Pl. Or. Nov. 1(5): 14 (1844); Boissier, Fl. Or. 4: 548 (1879); Post & Dinsmore, Fl. Syr. Pales. Sin. 2: 332 (1933); Thiebaut, Fl. Libano-Syr. 3: 45 (1953). *Amaracus libanoticus* (Boissier) Briquet, in Engler & Prantl, Nat. Pflanzenfam. 4(3a): 305 (1895). – Type: *Aucher 1656bis*, Lebanon, Lebanon Mts. (holo. G, iso. BM, FI, P).

Subshrubs, flowers bisexual. Roots c. 0.5 cm in diameter. Young shoots sparsely scabrous. Stems ± erect, up to 60 cm long, yellowish or purplish brown, slightly scabrous at the bases (hairs c. 0.1 mm long). Branches of the first order present, in the upper $\frac{1}{2}$ of the stems, up to 9 pairs per stem, 4 (1.5 – 8) cm long; sometimes branches of the second order present. *Leaves* up to 15 pairs per stem, more or less petiolate (petioles up to 8 mm long), roundish to ovate, tops usually obtuse, 11 (4 – 19) mm long, 8 (3 – 17) mm wide, more or less thin, light green, sometimes glaucous, margins scabrous (hairs c. 0.1 mm long), otherwise (nearly) glabrous, sessile glands inconspicuous, up to 600 per cm^2. *Spikes* (sub)globose, ovoid or cylindrical, 20 (15 – 40) mm long, 16 (10 – 23) mm wide, nodding. *Bracts* 8 (4 – 15) pairs per spike, ovate or oval, tops acute or acuminate, 9 (7 – 12) mm long, 7 (4 – 9) mm wide, usually vividly purple, margins scabrous, otherwise glabrous. *Calyces* 2-lipped for c. $\frac{2}{5}$, 5 (4 – 6) mm long, throats pilose, for the rest sparsely scabrous; upper lips divided, for c. $\frac{3}{5}$ (but varying), into 3, usually somewhat unequal, more or less triangular, 1.3 (0.9 – 2.1) mm long teeth; lower lips approximately as long as the upper lips, consisting of 2 (sub)equal (narrowly) triangular, 1.3 (0.9 – 2.1) mm long teeth. *Corollas* 2-lipped for c. $\frac{1}{3}$, 10 (8 – 13) mm long, pink, not saccate, outside sparsely pilosellous; upper lips divided, for c. $\frac{2}{5}$ (but varying), into 2, 0.7 (0.5 – 2.0) mm long lobes; lower lips divided, for c. $\frac{3}{5}$, into 3 (sub)equal, 2.3 (1.5 – 3.1) mm long lobes. *Stamens* ascending under the upper lips; filaments up to 9 and 11 mm long. *Styles* between the filaments protruding under the upper lips, up to 16 mm long.

G e o g r a p h y a n d e c o l o g y. *O. libanoticum* is an endemic species of the Lebanon Mts., where it has been found in several places, thus forming the southeastern border of the distribution area of the section *Anatolicon*. It grows from 700 – 2000 m, in dry regions and flowers from June to October.

N o t e s. 1. *O. libanoticum* differs from somewhat related species like *O. hyper-*

icifolium and *O. sipyleum* in its long calyx lower lips. From *O. scabrum* it differs in its ± ovate, often petiolate leaves. 2. In nature *O. libanoticum* forms now and then hybrids with *O. syriacum* (see p. 133).

LEBANON: Becharré, 1822, *Ehrenberg s.n.* (L, W). Lebanon Mts., 1837, *Aucher 1656bis* (type). Above Eden, June 1846, *Boissier s.n.* (E, G, K, L, P, W). Mar Tserkis, in shady places, c. 1500 m, 19 July 1855, *Kotschy 256* (BM, P, W). Mountains w. of Duma, 15 July 1868, *Post s.n.* (K). Al Farat, 17 Aug. 1898, *Post s.n.* (B). Lebanon Mts., 1899, *Bertrand s.n.* (G). Above Eden, on slopes, 1600 – 1800 m, 20 June 1910, *Bornmüller 12321* (B). Cedars, rocky places, 1 Sept. 1928, *Dinsmore 13929* (K). Near the monastery of Becharré, c. 1300 – 1400 m, 20 June 1931, *Zerny s.n.* (W). Baynu to Qornat Aruba, c. 800 m, 15 June 1943, *Davis 6315* (E). Cedars, Aug. 1944, *Gathorne-Hardy s.n.* (K).

11. **Origanum scabrum** Boissier et Heldreich – **Figs. 7, 8** and **9.**

O. scabrum Boissier et Heldreich, in Boissier, Diagn. Pl. Or. Nov. 1(7): 48 (1846); Boissier, Fl. Or. 4: 549 (1879). Halácsy, Consp. Fl. Graec. 2: 553 (1902); Ietswaart, Fokkinga & Vroman, Acta Bot. Neerl. 21: 439 (1972); Tutin et al., Fl. Eur. 3: 172 (1972). *Amaracus scaber* (Boissier et Heldreich) Briquet, in Engler & Prantl, Nat. Pflanzenfam. 4 (3a): 306 (1895). *O. scabrum* Boissier et Heldreich ssp. *euscabrum* (Hayek) Davis, Kew Bull. 1949: 405 (1949). *Amaracus scaber* (Boissier et Heldreich) Briquet ssp. *euscaber* Hayek, Prodr. Fl. Penins. Balc. 2: 332 (1931). – Type: *Heldreich s.n.*, Greece, Mt. Taygetos (holo. G, iso. BM, FI, L, W, WU).

O. pulchrum Boissier et Heldreich, in Boissier, Diagn. Pl. Or. Nov. 2(4): 11 (1859); Boissier, Fl. Or. 4: 549 (1879); Rechinger, Bot. Jahrb. 80: 394 (1961). *Amaracus pulcher* (Boissier et Heldrech) Briquet, in Engler & Prantl, Nat. Pflanzenfam. 4(3a): 306 (1895); Rechinger, Fl. Aegaea: 531 (1943). *O. scabrum* Boissier et Heldreich ssp. *pulchrum* (Boissier et Heldreich) Davis, Kew Bull. 1949: 405 (1949). *Amaracus scaber* (Boissier et Heldreich) Briquet ssp. *pulcher* (Boissier et Heldreich) Hayek, Prodr. Fl. Penins. Balc. 2: 332 (1932); Jackson, in Hooker's Ic. Pl. 34: 1 (1939). Type: *Heldreich s.n.*, Greece, Euboea, Mt. Delphi (holo. G, iso. G, JE, L, W).

Subshrubs, flowers bisexual, sometimes female only. Roots up to 1.5 cm in diameter. Young shoots scabrous. Stems erect or ascending and rooting at the bases, up to 45 cm long, light or dark brown, glabrous or slightly scabrous at the bases (hairs c. 0.1 mm long). Branches of the first order present, in the upper $\frac{2}{5}$ of the stems, up to 8 pairs per stem, 2.1 (0.4 – 6) cm long, not ramified. *Leaves* up to 12 pairs per stem, sessile, heart-shaped or roundish, tops acute or obtuse, 18 (5 – 33) mm long, 14 (3 – 24) mm wide, more or less leathery and glaucous, margins and veins beneath sparsely scabrous (hairs c. 0.1 mm long), otherwise glabrous, sessile glands up to 750 per cm^2. *Spikes* ovoid or subglobose, 18 (10 – 25) mm long, 14 (10 – 17) mm wide, nodding. *Bracts* 7 (3 – 12) pairs per spike, heart-shaped or ovate, tops acute or acuminate, 10 (7 – 13) mm long, 7 (5 – 10) mm wide, partly vividly purple, somewhat pilosellous. *Calyces* 2-lipped for c. $\frac{2}{5}$, (4 – 7) mm long, throats pilose, otherwise somewhat pilosellous; upper lips divided, for c. $\frac{2}{5}$ (but varying), into 3 (sub)equal, triangular or ± deltoid, 0.9 (0.1 – 2.5) mm long teeth; lower lips approximately as long as to clearly shorter than the upper lips, consisting of 2 (sub)equal, triangular, 1.4 (0.6 – 2.5) mm long teeth. *Corollas* 2-lipped for c. $\frac{1}{3}$, 11 (7 – 14) mm long, pink to purple, not saccate, outside somewhat pilosellous; upper lips divided, for c. $\frac{1}{5}$, into 2 0.5 (0.1 – 1.5) mm long lobes; lower lips divided, for c. $\frac{3}{5}$, into 3 subequal, 2.4 (2.0 – 3.5) mm long lobes. *Stamens* ascending under the upper lip; filaments up to 8 and 11 mm long. *Styles* between the filaments, protruding under the upper lips, up to 16 mm long.

Geography and ecology. *O. scabrum* is an endemic species of Greece, where it is known from four mountains: Mt. Taygetos and Mt. Malevo in the Peloponnisos, Mt. Kandilion and Mt. Delphi (including Mt. Xerobuni) on the island of Euboea. These sites constitute the northwestern limit of the area of the section *Anatolicon*. There is one herbarium specimen (under the name *O. pulchrum* in BM) from Kriti, collected near Nida in 1899 by Baldacci. This record is probably erroneous, owing to mislabelling. On the mountains mentioned above *O. scabrum* always occurs on limestone, in screes, between boulders and in cracks. Mostly it is found at altitudes between 1000 and 1800 m. Locally it is common. It flowers from June to September. For further ecological data see Ietswaart, Fokkinga & Vroman (1972).
Notes. 1. In the publication just mentioned it is shown that Boissier & Heldreich's *O. scabrum* and *O. pulchrum* are conspecific. 2. *O. scabrum* differs from *O. hypericifolium*, *O. libanoticum* and *O. sipyleum* in its larger, heart-shaped, sessile leaves. 3. Locally, *O. scabrum* forms a hybrid with *O. vulgare* ssp. *hirtum* (see p. 137), which has usually been considered as a good species and named *O. lirium*.

In the paper (loc. cit.) dealing with the delimitation of *O. scabrum* a summary has been given of the specimens studied. (These are included in the index of collections (II.8).)

12. **Origanum sipyleum** Linnaeus – **Figs. 7, 8** and **9.**

O. sipyleum Linnaeus, Sp. Pl.: 589 (1753); Sibthorp & Smith, Fl. Graeca 6: 57 (1826); Boissier, Fl. Or. 4: 547 (1879). *Majorana sipylea* (Linnaeus) Kosteletzky, Alg. Med.-Pharm. Fl. 3: 770 (1834). *Amaracus sipyleus* (Linnaeus) Rafinesque, Fl. Tell. 3: 86 (1836); Rechinger, Fl. Aegaea: 531 (1943); Wolf, Baileya 2: 60 (1954). – Type: *Linnaeus 743.3* (holo. LINN).

Subshrubs, flowers bisexual. Roots up to 1 cm in diameter. Young shoots tomentellous. Stems erect or ascending and rooting at the bases, up to 80 cm long, yellow or purplish brown, basely somewhat tomentellous (hairs c. 0.3 mm long), otherwise glabrous. Branches of the first order present, in the upper $\frac{1}{2}-\frac{3}{5}$ of the stems, up to 26 pairs per stem, 3 (0.5 – 35) cm long, usually with several pairs of small leaves; branches of the second order sometimes present. *Leaves* up to 32 pairs per stem, the lower ones petiolate (petioles up to 6 mm long), (longly) ovate, heart-shaped or (longly) oval, tops acute or obtuse, 10 (3 – 24) mm long, 6 (3 – 15) mm wide, more or less leathery, usually glaucous, glabrous, sessile glands inconspicuous, up to 500 per cm^2. *Spikes* subglobose or ovoid, seldom cylindrical, 14 (7 – 28) mm long, 8 (5 – 12) mm wide, more or less nodding. *Bracts* 9 (4 – 16) pairs per spike, obovate to oval, tops $\pm$ acute, 7 (4 – 10) mm long, 4 (3 – 6) mm wide, partly purple, glabrous. *Calyces* 2-lipped for c. $\frac{1}{5}$, 4 (4 – 6) mm long, throats pilose, for the rest glabrous; upper lips divided, for c. $\frac{2}{5}$, into 3 (sub)equal, $\pm$ deltoid, 0.4 (0.1 – 0.7) mm long teeth; lower lips somewhat shorter than the upper lips, consisting of 2 (sub)equal, $\pm$ deltoid, 0.3 (0.1

– 0.6) mm long teeth. *Corollas* 2-lipped for c. $\frac{1}{3}$, 9 (7 – 11) mm long, pink, (nearly) not saccate, outside sparsely pilosellous; upper lips divided for c. $\frac{1}{5}$ (but varying), into 2, 0.6 (0.1 – 1.1) mm long lobes; lower lips divided, for c. $\frac{3}{5}$, into 3 subequal, 2.6 (1.8 – 3.5) mm long lobes. *Stamens* sticking straight out or ascending under the upper lips; filaments up to 10 and 11 mm long. *Styles* sticking straight out, up to 14 mm long.

Geography and ecology. *O. sipyleum* has the largest distribution area of all species in the section *Anatolicon.* It has been found in the whole western half of Turkey, especially in the extreme west and south. Its most western site is on the island of Samos (see also note 2). It has been found from 100 – 1500 m, on calcareous rocks and rocky slopes, sometimes under pine trees. In one case it has been found in the steppes. It flowers from June to August.

Notes. 1. *O. sipyleum* differs from *O. libanoticum* and *O. scabrum* in its smaller calyx teeth, from *O. hypericifolium* in its less glandular punctate leaves. 2. Sibthorp & Smith (1826) mentioned *O. sipyleum* for Euboia. Possibly they mistook *O. scabrum (O. pulchrum)* for it. 3. *O. sipyleum* is supposed probably rightly, to be one of the parental species of the old garden hybrid *O.* × *hybridinum* (see p. 136). Furthermore *O. sipyleum* forms hybrids in natural sites with *O. onites* (p. 137). A third hybrid with *O. vulgare* ssp. *hirtum*, also found in nature, is putative (p. 142).

TURKEY. PROV. KONYA: Sultan Dağ near Akşehir, on rocks, 12 July 1899, *Bornmüller 5463* (B, G, JE, WU). Valley of Tekké near Akşehir, 7 July 1907, *Saint-Lager s.n.* (G). Sultandağ, in rocky steppe and on slope with debris, c. 1400 m, 19 July 1968, *Sorger 68 – 46 – 40* (Herb. Sorger). PROV. ANTALYA: Tahtalidağ, above Burnarbasi, between shrubs, on limestone, 200 – 500 m, 6 July 1933, *Schwarz 792* (B). PROV. ISPARTA: on sandy hills, 1845, *Heldreich s.n.* (G, L, W). PROV. DENIZLI: Tas Ocaği near Denizli, on rocky places, 13 July 1947, *Davis 13256* (B, E, G). PROV. MUGLA: 1843, *Pinard s.n.* (G). PROV. AYDIN: 25 June 1886, *Forsyth Major 713* (G). PROV. IZMIR: near Izmir, 1870, *Peronin s.n.* (G). Near Ephesos, c. 100 m, 1 July 1967, *Sorger 67 – 44 – 12* (Herb. Sorger). PROV. MANISA: Mt. Sipylus, June 1842, *Boissier s.n.* (G, JE, L). Mt. Sipylus near Magnésie, 14 Aug. 1854, *Balansa 328* (G, JE). Mt. Sipylus above Magnésia, between shrubs, 12 Aug. 1933, *Schwarz 965* (B). C. 27 km s. of Demirçi, in Quercetum, on limestone, c. 540 m, 23 June 1954, *Huber-Morath 12707* (Herb. Huber-Morath). C. 5 km s.e. of Soma, in *Quercus* macchie, c. 140 m, 25 June 1964, *Huber-Morath 17209* (Herb. Huber-Morath). PROV. ESKIŞEHIR: n. of Ouchak, stony hills, c. 950 m, 13 June 1857, *Balansa 1174* (G, JE). PROV. ÇANKIRI: near Eldivan, along the road to Şabanözü, on rocks, c. 1080 m, 20 Aug. 1971, *Buttler 15529* (Herb. Buttler). PROV. KASTAMONU (?): on hills, July and Aug. 1892, *Sintenis 4911* (B, COI, G, JE, L).

GREECE. SAMOS: between Karlowasi and Maralliokampsos, on calcareous rocks, 16 – 23 June 1932, *Rechinger 2023* (W).

13. **Origanum vetteri** Briquet et Barbey – **Figs. 7, 8** and **9.**

O. vetteri Briquet et Barbey, in de Stefani, Forsyth Major & Barbey, Karpathos, Étude Géol. Pal. Bot.: 124 (1895); Tutin et al., Fl. Eur. 3: 172 (1972). *Amaracus vetteri* (Briquet et Barbey) Briquet, in Engler & Prantl, Nat. Pflanzenfam. 4(3a): 306 (1895); Hayek, Prodr. Fl. Penins. Balc. 2: 332 (1931); Rechinger, Fl. Aegaea: 531 (1943). – Type: *Forsyth Major 262*, Greece, Karpathos, Mt. Kalolimni (holo. G, iso. COI, FI, K, P).

Caespitose subshrubs, flowers bisexual. Roots up to 1.5 cm in diameter. Young shoots lanata-pilose. Stems many together in one specimen, ascending and rooting at the bases, up to 10 cm long, brown, lanato-pilose (hairs c. 1 mm long). Branches of the first order usually absent, but sometimes present at the top of the stems, up to 2 pairs per stem, c. 1 cm long, not ramified. *Leaves* up to 11 pairs per stem, shortly petiolate (petioles up to 5 mm long), heart-shaped, ovate or somewhat deltoid, tops more or less obtuse, 4 (2 – 8) mm long, 3 (1 – 5) mm wide, thin, light green, lanato-pilose (hairs c. 1 mm long), sessile glands inconspicuous, up to 500 per cm^2; margins revolute; veins more or less raised at the underside. *Spikes* subglobose 8 (6 – 11) mm long, 10 (8 – 12) mm wide, slightly nodding. *Bracts* 5 (3 – 7) pairs per spike, obovate to oval, tops acute or acuminate, 7 (5 – 8) mm long, 4 (3 – 5) mm wide, partly purplish, pilose. *Calyces* 2-lipped for c. $\frac{2}{5}$, 5 (4 – 5) mm long, throats and otherwise pilose; upper lips divided, for c. $\frac{1}{2}$ (but varying), into 3 (sub)equal, deltoid or triangular, 1.0 (0.7 – 1.5) mm long teeth; lower lips somewhat shorter than the upper lips, consisting of 2 (sub)equal, acuminate, triangular, 1.3 (0.9 – 1.7) mm long teeth. *Corollas* 2-lipped for c. $\frac{1}{3}$, 7 (5 – 9) mm long, pink, not saccate, outside somewhat pilosellous; upper lips divided for c. $\frac{2}{5}$, into 2, 0.6 (0.4 – 1.1) mm long lobes; lower lips divided for c. $\frac{2}{5}$, into 3 slightly unequal, 1.4 (0.6 – 2.0) mm long lobes. *Stamens* sticking straight out; filaments up to 4 and 5 mm long. *Styles* sticking straight out, up to 12 mm long.

Geography and ecology. *O. vetteri* is an endemic species of the island of Karpathos, where it is found on Mt. Kalolimni and Mt. Kolla, at c. 1100 m, on calcareous rocks, in fissures. It has been found flowering in June and July.

Notes. 1. *O. vetteri* possesses several characters which are uncommon in the genus *Origanum*: a dwarf, caespitose habit, very small lanato-pilose leaves with revolute margins and somewhat raised veins, frequently occurrence of terminal spikes only. Several of these characters are more commonly found in the genus *Thymus*. 2. From all other species in the section *Anatolicon*, *O. vetteri* differs in its dwarf habit and very small leaves.

GREECE. KARPATHOS: at the top of Mt. Kalolimni, May 1883, *Pichler 526* (G). Mt. Kalolimni, in fissures in calcareous rocks near the top, 8 July 1886, *Forsyth Major 262* (type). Mt. Kalolimni, on calcareous rocks, c. 1100 m, 15 June 1935, *Rechinger 8187* (B, G, JE, K, W). Mt. Kalolimni, n.w. side, among rocks at foot of cliff, growing with mosses, c. 1100 – 1200 m, 21 July 1950, *Davis 18005* (E).

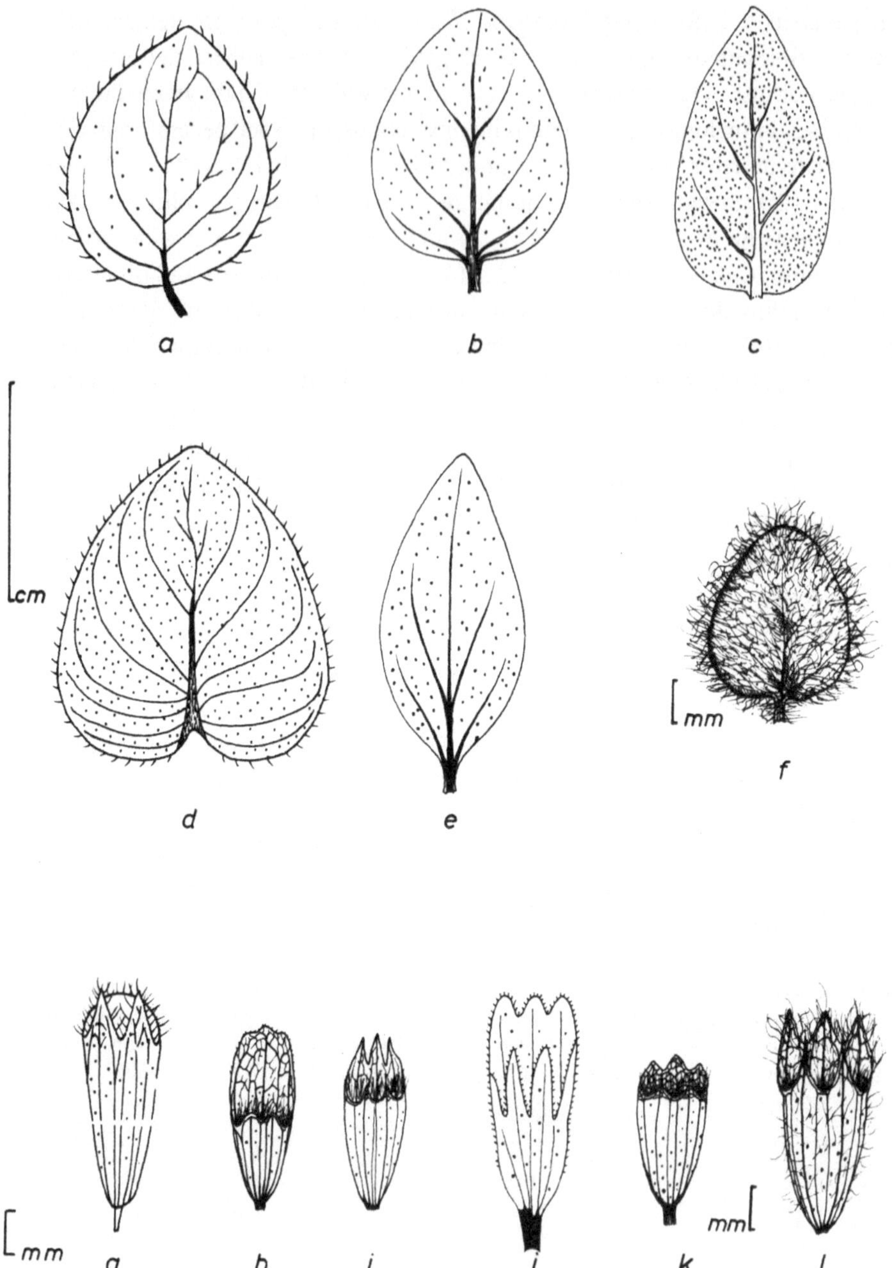

Figure 7. Leaves and calyces of the species in the section *Anatolicon*, except *O. libanoticum* (for which see figure 10): a. and g. *O. akhdarense;* b. and h. *O. cyrenaicum;* c. and i. *O. hypericifolium;* d. and j. *O. scabrum;* e. and k. *O. sipyleum;* f. and l. *O. vetteri.*

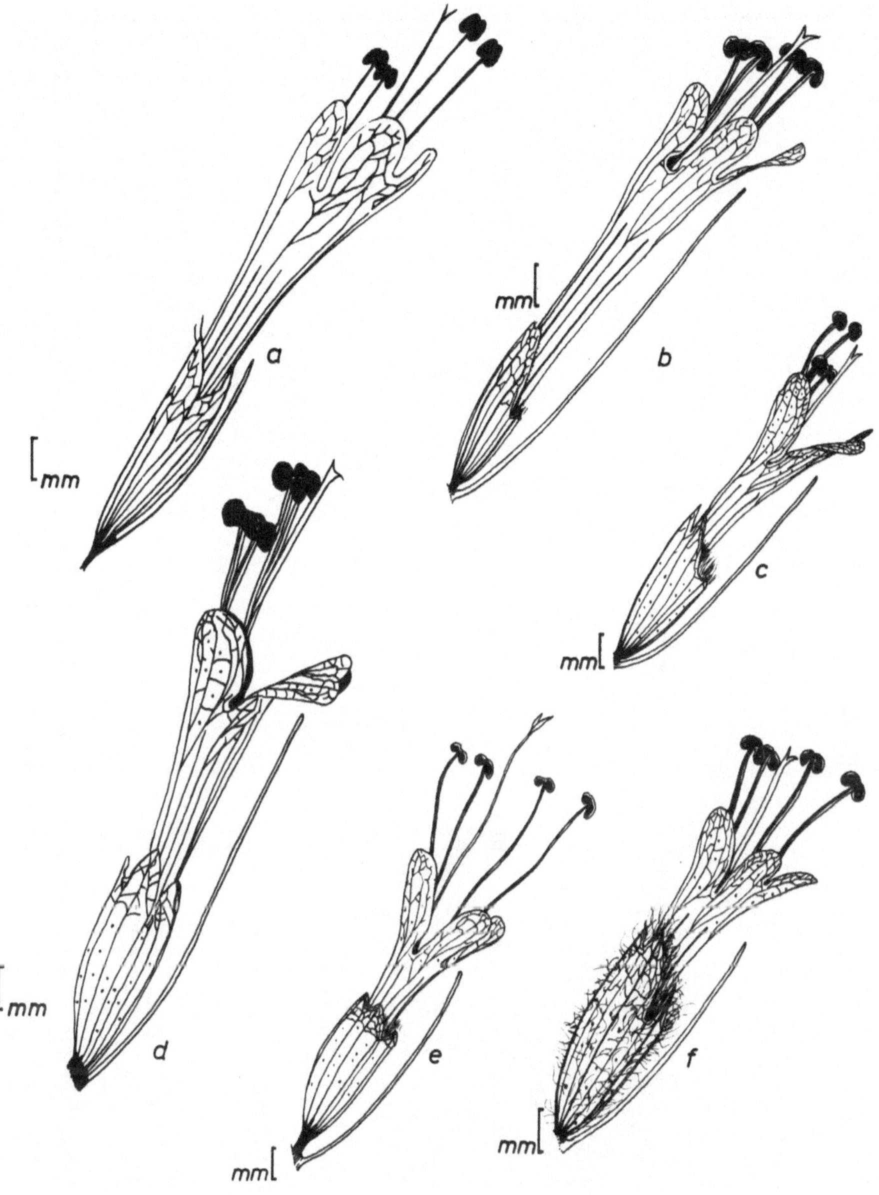

Figure 8. Flowers with bracts in side view of the species in the section *Anatolicon*, except *O. libanoticum* (for which see figure 10): a. *O. akhdarense;* b. *O. cyrenaicum;* c. *O. hypericifolium;* d. *O. scabrum;* e. *O. sipyleum;* f. *O. vetteri.*

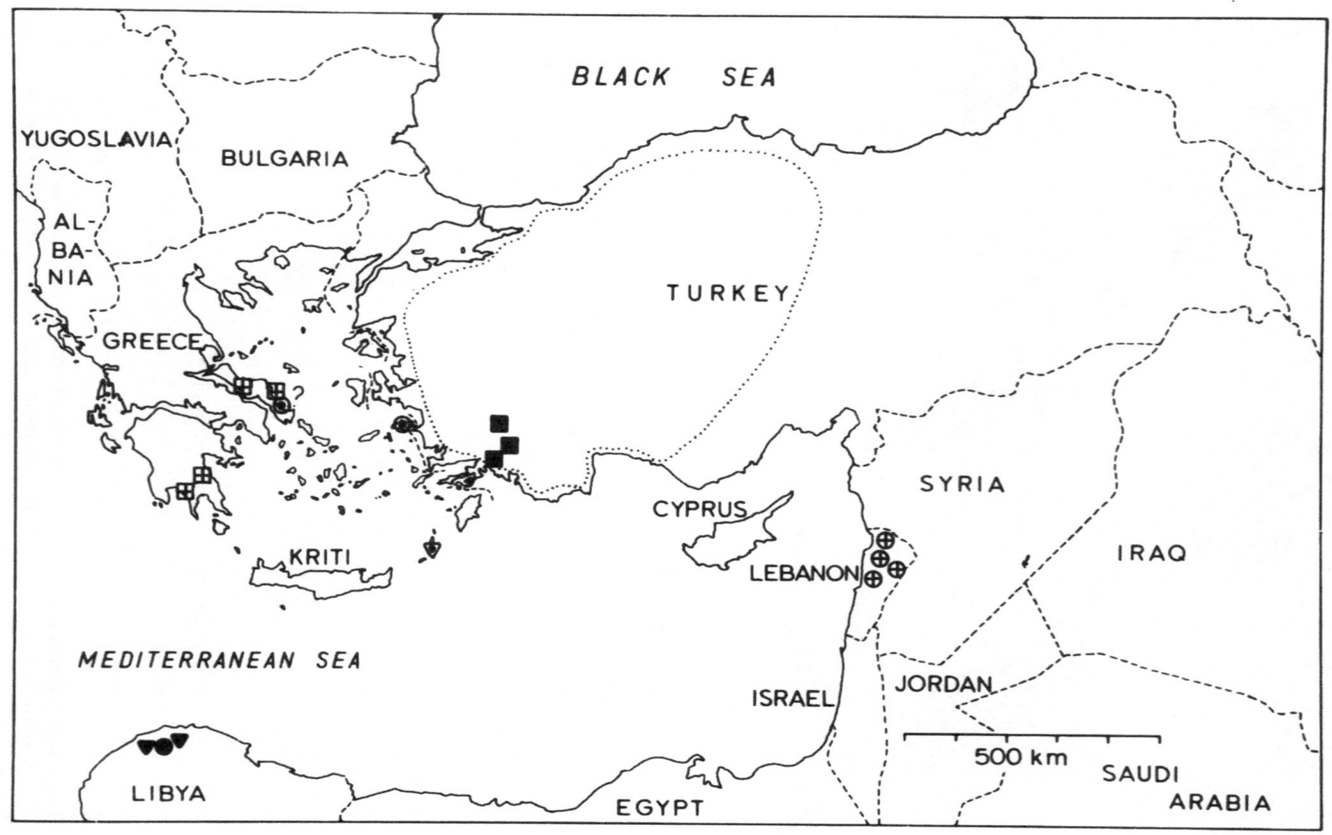

Figure 9. Distribution of the species in the section *Anatolicon*: ● *O. akhdarense*; ▼ *O. cyrenaicum*; ■ *O. hypericifolium*; ⊕ *O. libanoticum*; ⊞ *O. scabrum*; . . . . and ⊙ *O. sipyleum*; ▽ *O. vetteri*; '? based on one doubtful record.

Figure 10. O. libanoticum: a. habit; b. leaf; c. bract with calyx in upper lip view; d. calyx in lower lip view; e. calyx cut through the lower lip; f. flower with bract in side view; g. corolla cut through the lower lip.

III. Section **Brevifilamentum** Ietswaart

Section *Brevifilamentum* Ietswaart, Not. R. B. G. Edinburgh 38(1): 46 (1980). – Type: *Origanum rotundifolium* Boissier.

Branches of the first order usually present, those of the second order seldom so. Leaves mostly more or less leathery. *Spikes* normally large and nodding. *Bracts* imbricate, $1\frac{1}{5}$ – 3 x calyces, usually membranous and glabrous. *Flowers* usually several per verticillaster and shortly pedicellate, bisexual, large. *Calyces* tubular, 2-lipped for c. $\frac{2}{5}$; teeth in upper and lower lips well developed; throats always pilose. *Corollas* 2-lipped for c. $\frac{1}{5}$, c. 2 x calyces, not saccate. *Stamens* very unequal in length, the upper 2 very short and included, the lower 2 long, or more or less short, ascending under the upper lips and either slightly or far protruding; filaments c. $\frac{1}{10}$ respectively $\frac{1}{5}$ or $\frac{3}{5}$ × corollas.

14. **Origanum acutidens** (Handel-Mazzetti) Ietswaart – **Figs. 11, 12** and **13.**

O. acutidens (Handel-Mazzetti) Ietswaart, *stat. et comb. nov. Amaracus haussknechtii* (Boissier) Briquet var. *acutidens* Handel-Mazzetti, Ann. K.K. Naturh. Hofm. 27: 420 (1913); Bornmüller, Notizbl. Bot. Gart. Berlin 7(63): 26 (1917). – Type: *Handel-Mazzetti 2913*, Turkey, Bitlis, between Goro and Kede (holo. W, iso. WU).

Subshrubs. Roots up to 1 cm in diameter. Young shoots sparsely scabrous. Stems erect, up to 50 cm long, yellow or purplish brown, sparsely scabrous at the base (hairs c. 0.1 mm long), otherwise glabrous. Branches of the first order usually present, in the upper $\frac{2}{5}$ of the stems, up to 10 pairs per stem, 3 (0.5 – 15) cm long, usually not ramified. *Leaves* up to 16 pairs per stem, (sub)sessile, usually ovate, tops obtuse, 16 (5 – 30) mm long, 11 (4 – 24) mm wide, more or less leathery, both sides glaucous, ± glabrous, sessile glands numerous, up to 1800 per cm^2. *Spikes* subglobose, sometimes cylindrical, 22 (10 – 43) mm long, 21 (10 – 32) mm wide, nodding. *Bracts* 8 (3 – 12) pairs per spike, roundish, obovate or oval, tops more or less acute or obtuse, 13 (7 – 22) mm long, 13 (6 – 20) mm wide, yellowish green, glabrous. *Flowers* 4 (2 – 12) per verticillaster, with c. 1 mm long pedicels. *Calyces* 2-lipped for c. $\frac{2}{5}$, 6 (5 – 7.5) mm long, throats pilose, otherwise very sparsely pilosellous; upper lips divided, for c. $\frac{1}{2}$ (but varying), into 3 (sub)equal, ± deltoid, ovate or triangular, 1.1 (0.6 – 2.4) mm long teeth; lower lips slightly longer to slightly shorter than the upper lips, consisting of 2 (sub)equal, (narrowly) triangular, acute or acuminate, 2.3 (1.5 – 3.2) mm long teeth. *Corollas* 2-lipped for c. $\frac{1}{5}$, 13 (10 – 16) mm long, white, sometimes slightly pink, not saccate, outside pilosellous; upper lips divided, for c. $\frac{1}{10}$ (but very variable), into 2, 0.6 (0.2 – 1.4) mm long lobes; lower lips divided, for c. $\frac{1}{2}$, into 3 somewhat unequal, 1.6 (0.9 – 2.7) mm long lobes. *Stamens*, the upper 2 included, the lower 2 far protruding; filaments up to 2 and 10 mm long. *Styles* protruding under the upper lips, up to 22 mm long.

Geography and ecology. *O. acutidens* occurs in a rather large area in north-eastern Turkey. It is possible that it also occurs beyond the Turkish eastern border, but there are no records from that region until now. As far as is now known the distribution of *O. acutidens* does not overlap those of related species *O. haussknechtii* and *O. rotundifolium*. From the border of the first 2 species a few specimens have been seen with some characters intermediate. *O. acutidens* is found on limestone and on non-calcareous soils, from 1000 – 3000 m, sometimes in shady places. It flowers from June to August.

Note. *O. acutidens* has been originally described as a variety of *O. haussknechtii*. Although doubtless related to this species it differs from it in several characters: its somewhat larger, less leathery leaves, its usually subglobose spikes with fewer bracts, its larger, roundish, yellowish green bracts, its longer, acute or acuminate teeth in the calyx lower lips, and its white (or pinkish corollas). So a specific rank is justified. *O. acutidens* is also akin to *O. rotundifolium*, from which it differs in its slightly scabrous stems, its ovate leaves with much shorter hairs, more glandular punctations and less conspicuous veins, its somewhat smaller bracts and its smaller more acutely toothed calyces.

TURKEY. PROV. HAKKARI: Cilo Dağ, in Diz Deresi, rocky slopes in shade, c. 1950 m, 10 Aug. 1954, *Davis & Polunin 24283* (BM, E). Morinos Dere opposite Marunis, steep eroded calcareous slope, 1550 m, 21 June 1966, *Davis 45325* (E, K). PROV. BITLIS: valley of Sassun, between Goro and Kede, on calcareous rocks, c. 1200 – 1600 m, 12 Aug. 1910, *Handel-Mazzetti 2913* (type). PROV. MALATYA: near Malatya, July 1936, *Gleisberg 205* (B). PROV. SIVAS: c. 18 km s. of Zara, c. 1500 m, 9 July 1969, *Sorger 69 – 37 – 31* (Herb. Sorger). PROV. TUNÇELI: Pertek Hozat, eroded shaley hills, c. 1600 m, 13 July 1957, *Davis & Hedge 31067* (E). C. 27 km n. of Tunçeli, in Quercetum, c. 960 m, 7 July 1959, *Huber-Morath 15267* (Herb. Huber-Morath). PROV. ERZINCAN: Kurutschai near Nerzkiep, on rocky slopes, 28 June 1889, *Sintenis 1027* (P, WU). Erzinghan, Bogos Chan, near the Euphraat, on rocky slopes, 17 July 1890, *Sintenis 2996* (E, G, JE). Sipikov, Kainik Dere, on rocky slopes, 30 July 1890, *Sintenis 3252* (B, JE, P). Keşiş Dağ above Cimin, on igneous scree, c. 2800 m, 27 July 1957, *Davis & Hedge 31637* (BM, K). Keşiş Dağ above Cimin, rocky metamorphic slopes, 1900 – 2100 m, 26 July 1957, *Davis & Hedge 31771* (BM, K). PROV. ERZERUM: gorge between Tercan and Selcpur, shaley banks, 11 July 1957, *Davis & Hedge 30953* (BM, W). C. 70 km from Hinis to Erzerum, bare stony banks in Aras gorge, c. 1600 m, 12 July 1966, *Davis 46451* (E, K). PROV. GÜMÜSANE: Szandschak Gümüschkhane, Taltaban, on rocky slopes, 28 Aug. 1894, *Sintenis 6102* (E, G, JE, L). Gümüsh Hane, Karshut-su, c. 1100 m, 27 July 1930, *Frödin s.n.* (W). Gümüsh Hane, steep non-lime screes, 17 June 1934, *Balls 1707* (E, G).

15. **Origanum bargyli** Mouterde – **Figs. 11, 12** and **13.**

O. bargyli Mouterde, Saussurea 4: 22 (1973). – Type: *Joseph Louis s.n.*, Syria, Slenfé (holo. P, iso. G).
O. brevidens (Bornmüller) Dinsmore var. *pubescens* Thiebaut, Fl. Libano-Syr. 3: 46 (1953).

Subshrubs. Young shoots hirtellous. Stems up to 40 cm long, ascending and rooting at the bases, light or purplish brown, somewhat hirtellous at the bases (hairs c. 0.6 mm long), otherwise glabrous. Branches of the first order absent or a few present, in the upper $\frac{2}{5}$ of the stems, up to 4 cm long, not ramified. *Leaves* up to 16 pairs per stem, (sub)sessile, heart-shaped or ovate, tops obtuse to acuminate, 12 (4 – 19) mm long, 10 (4 – 17) mm wide, rather thin, light green or purplish, more or less glaucous, sparingly hirtellous or scabrous at least at the margins (hairs c. 0.3 mm long), sessile glands up to 500 per cm^2. *Spikes* subglobose, ovoid or cylindrical, 20 (10 – 25) mm long, 12 (9 – 15) mm wide, more or less nodding. *Bracts* 7 (4 – 14) pairs per spike, ovate, oval or obovate, tops usually acuminate, 9 (5 – 17) mm long, 7 (3 – 13) mm wide, partially purple, margins pilosellous. *Flowers* 2 (2 – 6) per verticillaster, with c. 1 mm long pedicels. *Calyces* 2-lipped for c. $\frac{2}{5}$, 6 (5 – 8) mm long, throats pilose, margins pilosellous; upper lips divided, for c. $\frac{3}{5}$ (but varying), into 3 (sub)equal, ± triangular, 1.5 (1.0 – 2.1) mm long teeth; lower lips approximately as long as the upper lips, consisting of 2 (sub)equal, triangular, 2 (1.5 – 3.0) mm long teeth. *Corollas* 2-lipped for c. $\frac{1}{5}$, 13 (11 – 16) mm long, pink, not saccate, outside pilosellous; tubes slightly curved downwards, upper lips divided, for c. $\frac{1}{5}$, into 2 c. 0.5 mm long lobes; lower lips divided, for c. $\frac{1}{2}$, into 3 somewhat unequal, c. 1.5 mm long lobes. *Stamens*, the upper 2 included, the lower 2 shortly protruding; filaments c. 1 and 3 mm long. *Styles* protruding under the upper lips, up to 17 mm long.

Geography and ecology. The two places in which *O. bargyli* has been found until now are both mountain areas (c. 1200 m). In one it grows in an open pine wood. Notes. 1. The type specimens from Slenfé differ in some respects from the specimens collected near Osmaniye: in all parts they are smaller, in addition the leaves are obtuse and the spikes more or less cylindrical. In characters of calyces and corollas they do not differ. 2. Thiebaut (1953) regarded the taxon in question as a variety of *O. brevidens*. Mouterde considered it, rightly, as a species. Undoubtedly *O. bargyli* is related to *O. brevidens*, but it differs from it in several characters: its hirtellous to scabrous stems and leaves, its thinner leaves and its somewhat larger calyces with longer teeth. In some characters (e.g. indumentum, and shape of leaves, spikes, bracts and calyces) *O. bargyli* is reminiscent of *O. amanum*. The main differences between the two species lie in the corollas, which are in the latter species much longer and in the possession of four subsessile, included stamens. All three species are more or less sympatric. More specimens need to become available for a better understanding of their relationship. 3. The type of corollas found in *O. bargyli* and *O. brevidens* also occurs in *O. leptocladum*. 4. From the type locality a hybrid is known between *O. bargyli* and *O. syriacum* (see p. 139).

SYRIA: Jebel el Ansariya, Slenfé, 12 July 1939, *Joseph Louis s.n.* (type).
TURKEY. PROV. SEYHAN: near Yaglipinar, above Yarpuz, in Pinetum, c. 1250 m, 2 July 1959, *Huber-Morath 15266* (Herb. Huber-Morath).

16. **Origanum brevidens** (Bornmüller) Dinsmore – **Figs. 11, 12** and **13.**

O. brevidens (Bornmüller) Dinsmore, in Post & Dinsmore, Fl. Syr. Palest. Sin. 2: 332 (1933). Thiebaut, Fl. Libano-Syr. 3: 46 (1953). *Amaracus brevidens* Bornmüller, Notizbl. Bot. Gart. Berlin 7(63): 26 (1917). – Type: *Meinke 113*, Turkey, Hatay, Amanus Mts. (holo. B).

Subshrubs. Stems ascending or erect, c. 20 cm long, purplish or dark brown, glabrous. Branches of the first order often absent, but not necessarily, when present, in the upper part of the stems c. 1.5 cm long, not ramified. *Leaves* sessile, heart-shaped or ovate, tops ± acute, c. 16 mm long, c. 14 mm wide, more or less leathery, glabrous, sessile glands c. 600 per cm^2. *Spikes* subglobose or cylindrical, c. 25 mm long, throats pilose otherwise glabrous; upper lips divided, for c. $\frac{2}{3}$, into 3 (sub)-equal, deltoid, c. 0.7 mm long teeth; lower lips approximately as long as the upper glabrous. *Flowers* 2 per verticillaster, subsessile. *Calyces* 2-lipped for c. $\frac{2}{5}$, c. 5 mm long, throats pilose otherwise glabrous; upper lips divided, for c. $\frac{2}{5}$, into 3 (sub)-equal, deltoid, c. 0.7 mm long teeth; lower lips approximately as long as the upper lips, consisting of 2 (sub)equal, triangular, c. 1.5 mm long teeth. *Corollas* 2-lipped for c. $\frac{1}{5}$, c. 14 mm long, pink, not saccate, outside pilosellous; tubes slightly curved downwards; upper lips divided, for c. $\frac{1}{5}$, into 2, c. 0.5 mm long lobes; lower lips divided, for c. $\frac{1}{2}$, into 3, slightly unequal, c. 1.3 mm long lobes. *Stamens*, the upper 2 included, the lower 2 shortly protruding; filaments c. 1 and 4 mm long. *Styles* protruding under the upper lips, c. 16 mm long.

Geography and ecology. *O. brevidens* has been found only once in the Amanus Mts. Further data are not known.

Notes. 1. Very scanty herbarium material, of the type specimen only, was available for study, so *O. brevidens* is also in this respect imperfectly known. 2. *O. brevidens* is related to *O. bargyli*, from which it differs in its glabrous stems, leaves and bracts, its more leathery leaves and smaller, less toothed calyces (see also notes for *O. bargyli*).

TURKEY. PROV. HATAY: Amanus Mts., 1000 – 1700 m, 1909 – 1910, *Meinke 113* (type).

17. Origanum haussknechtii Boissier – Figs. 11, 12 and 13.

O. haussknechtii Boissier, Fl. Or. 4: 550 (1879). *Amaracus haussknechtii* (Boissier) Briquet, in Engler & Prantl, Nat. Pflanzenfam. 4(3a): 306 (1895); Bornmüller, Notizbl. Bot. Gart. Berlin 7(63): 27 (1917). – Type: *Haussknecht s.n.*, Turkey, Adiyaman, Mt. Akdagh (holo. G, iso. JE).

Subshrubs. Roots up to 1.5 cm in diameter. Young shoots hirtellous. Stems erect, up to 50 cm long, light or purplish brown, sparsely hirtellous at the bases (hairs c. 0.3 mm long), otherwise glabrous. Branches of the first order present or absent, in the upper $\frac{1}{2}$ of the stems, up to 8 pairs per stem, 4 (2 – 30) cm long, not ramified. *Leaves* up to 14 pairs per stem, sessile, heart-shaped, tops obtuse or acute, 13 (5 – 23) mm long, 10 (4 – 17) mm wide, leathery, both sides glaucous, ± glabrous, sessile glands up to 1200 per cm^2. *Spikes* cylindrical, sometimes subglobose, 30 (15 – 45) mm long, 19 (14 – 25) mm wide, nodding. *Bracts* 10 (6 – 18) pairs per spike, obovate, roundish or oval, tops obtuse, 10 (6 – 14) mm long, 9 (5 – 13) mm wide, partly purple, glabrous. *Flowers* 3 (2 – 7) per verticillaster, with c. 1 mm long pedicels. *Calyces* 2-lipped for c. $\frac{2}{5}$, 6 (5 – 7) mm long, throats pilose, otherwise glabrous; upper lips divided, for c. $\frac{2}{5}$, into 3 (sub)equal, somewhat ovate, obtuse 1.5 (0.9 – 1.8) mm long teeth; lower lips somewhat shorter than the upper lips, consisting of 2 (sub)equal, ovate or triangular, ± obtuse, 1.7 (1.4 – 2.0) mm long teeth. *Corollas* 2-lipped for c. $\frac{1}{5}$, 14 (12 – 16) mm long, pink, not saccate, outside pilosellous; upper lips divided, for c. $\frac{1}{10}$ (but varying), into 2, 0.4 (0.1 – 1.2) mm long lobes; lower lips divided, for c. $\frac{1}{2}$, into 3 somewhat unequal, 1.7 (1.0 – 3.1) mm long lobes. *Stamens*, the upper 2 included, the lower 2 far protruding; filaments up to 2 and 9 mm long. *Styles* protruding under the upper lips, up to 20 mm long.

Geography and ecology. *O. haussknechtii* has been collected on a few places in eastern Turkey. It grows in mountain regions from 1000 – 1650 m. It flowers from July to September.
Note. In its relatively small, ovate, purplish bracts, its obtuse toothed calyces with relatively short lower lips and its pink corollas *O. haussknechtii* differs from the related species, *O. acutidens* and *O. rotundifolium*.

TURKEY. PROV. ADIYAMAN: Mt. Akdagh above Adiaman and Malatia, c. 1650 m, 12 Sept. 1865, *Haussknecht s.n.* (type). PROV. MARAŞ: near Marasch, 1865, *Haussknecht s.n.* (JE). Cultivated specimens of the above collection (JE). PROV. MALATYA: c. 17 km s. of Kemaliya, c. 1000 m, 23 June 1949, *Huber-Morath 9394* (Herb. Huber-Morath).

18. **Origanum leptocladum** Boissier – **Figs. 11, 12** and **13.**

O. leptocladum Boissier, Fl. Or. 4: 549 (1879). *Amaracus leptocladus* (Boissier) Briquet, in Engler & Prantl, Nat. Pflanzenfam. 4(3a): 306 (1895). – Type: *Péronin s.n.*, Turkey, Konya, Yemourdaba Dagh (holo. G, iso. BM, JE, P).
Non *Majorana leptoclados* Rechinger, Denkschr. Akad. Wiss. Wien, Math.-Nat. Kl. 105: 125 (1943), (= *O.* × *minoanum* Davis).

Subshrubs. Roots up to 1 cm in diameter. Young shoots glabrous. Stems erect, up to 65 cm long, dark purplish brown, somewhat glaucous, glabrous. Branches of the first order usually present, in the upper $\frac{2}{5}$ of the stems, up to 10 pairs per stem, 6 (1.5 – 12) cm long, usually not ramified. *Leaves* up to 22 pairs per stem, sessile, heart-shaped or ovate, tops acute or apiculate, 12 (3 – 17) mm long, 9 (2 – 15) mm wide, leathery, both sides glaucous, glabrous, sessile glands up to 900 per cm^2. *Spikes* slender, cylindrical, 20 (8 – 45) mm long, 6 (4 – 8) mm wide, nodding. *Bracts* 10 (3 – 16) pairs per spike, more or less lanceolate, tops acuminate, 6 (5 – 8) mm long, 2 (1 – 3) mm wide, more or less leathery, partly purplish, often glaucous, glabrous. *Flowers* 2 per verticillaster, subsessile. *Calyces* 2-lipped for c. $\frac{2}{5}$, 5 (4 – 6) mm long, throats pilose, otherwise glabrous; upper lips divided, for c. $\frac{2}{5}$, into 3 (sub)equal, deltoid, 0.5 (0.2 – 0.8) mm long teeth; lower lips somewhat shorter than the upper lips, consisting of 2 (sub)equal, triangular, 0.8 (0.4 – 1.0) mm long teeth. *Corollas* 2-lipped for c. $\frac{1}{5}$, 12 (8 – 14) mm long, pink, not saccate, outside pilosellous; tubes usually slightly curved downwards; upper lips divided, for c. $\frac{1}{5}$ (but varying), into 2, 0.3 (0.2 – 0.4) mm long lobes; lower lips divided, for c. $\frac{1}{2}$, into 3 subequal, 1.0 (0.5 – 1.5) mm long lobes. *Stamens*, the upper 2 included, the lower 2 shortly protruding; filaments up to 1.5 and 5 mm long. *Styles* protruding under the upper lips, up to 20 mm long.

Geography and ecology. *O. leptocladum* has been found until now in a few places in southern Turkey, where it occurs on mountain slopes at a height of c. 1500 m, on chalk. It flowers in July and August.
Note. *O. leptocladum* is a remarkable species in the section *Brevifilamentum* through its very slender spikes, and its small bracts. In these characters and its glabrous habit it is reminiscent of *O. laevigatum* (section *Prolaticorolla*).

TURKEY. PROV. KONYA: Yemourdaba Dagh, near Ermenek, July 1872, *Péronin s.n.* (type). Ermenek, Balkusan deresi, on marly hillsides, 1500 m, 14 Aug. 1949, *Davis 16194* (E, G, K, W).

19. **Origanum rotundifolium** Boissier – **Figs. 13** and **14.**

O. rotundifolium Boissier, Diagn. Pl. Or. Nov. 2(4): 11 (1859); Boissier, Fl. Or. 4: 550 (1879). *Amaracus rotundifolius* (Boissier) Briquet, in Engler & Prantl, Nat. Pflanzenfam. 4(3a): 306 (1895); Handel-Mazzetti, Ann. K.K. Naturh. Hofm. 27: 420 (1913); Bornmüller, Notizbl. Bot. Gart. Berlin 7(63): 26 (1917). – Type: *Calvert s.n.*, Turkey, Erzerum, near Batum (holo. G).

Subshrubs. Roots up to 1 cm in diameter. Young shoots hirsute. Stems ascending and rooting at the bases or erect, up to 30 cm long, yellowish or purplish brown, more or less hirsute (hairs c. 1 mm long). Branches of the first order absent or a few present, in the upper $\frac{2}{5}$ of the stems, up to 5 pairs per stem (often only 1 at a node), 2.5 (1 – 5) cm long, not ramified. *Leaves* up to 17 pairs per stem, subsessile, usually heart-shaped or roundish, tops obtuse, 15 (6 – 25) mm long, 13 (4 – 20) mm wide, more or less leathery, light green, ± glaucous, sparsely hirtellous (hairs c. 0.5 mm long), sessile glands up to 400 per cm^2; veins conspicuous and usually raised at the under side. *Spikes* (sub)globose, sometimes cylindrical or pyramidate, 26 (12 – 60) mm long, 25 (18 – 37) mm wide, more or less nodding. *Bracts* 8 (3 – 17) pairs per spike, roundish, often wider than long, tops obtuse, 14 (8 – 24) mm long, 15 (7 – 27) mm wide, yellowish green, glabrous. *Flowers* 6 (2 – 16) per verticillaster, with c. 1 mm long pedicels. *Calyces* 2-lipped for c. $\frac{2}{5}$, 7 (5 – 9) mm long, throats pilose, otherwise glabrous; upper lips divided, for c. $\frac{2}{5}$ (but varying), into 3 (sub)equal, broadly ovate or ± deltoid, 1.3 (0.7 – 2.0) mm long teeth; lower lips slightly shorter than the upper lips, consisting of 2 (sub)equal, ovate or triangular, acute or ± obtuse, 2.4 (1.5 – 3.3) mm long teeth. *Corollas* 2-lipped for c. $\frac{1}{5}$, 13 (9 – 16) mm long, white or pale pink, not saccate, outside pilosellous; upper lips divided, for c. $\frac{1}{10}$ (but very variable), into 2, 0.4 (0.1 – 1.2) mm long lobes; lower lips divided, for c. $\frac{1}{2}$, into 3 slightly unequal, 1.8 (1.0 – 2.8) mm long lobes. *Stamens*, the upper 2 included, the lower 2 far protruding; filaments up to 2 and 10 mm long. *Styles* protruding under the upper lips, up to 21 mm long.

Geography and ecology. *O. rotundifolium* has a rather large distribution area in northeastern Turkey and the adjacent part of the U.S.S.R. Its area forms the northeastern limit of the section *Brevifilamentum*. It occurs in mountain regions, from 400 – 1500 m, and flowers from June to September.

Note. *O. rotundifolium* is related to *O. acutidens*, but it differs from this species in its hirsute stems, its shorter internodes, longer hairs on stems and leaves, its usually heart-shaped leaves with conspicuous veins and less glandular punctations, its somewhat larger spikes and its less acute calyx teeth.

TURKEY. PROV. KARS: near Salaczur, on rocks, 6 July 1931, *Koenig s.n.* (BM, WU). PROV. ÇORUH: Çoruh gorge between Artvin and Borçka, eroded igneous banks, 900 m, 21 June 1937, *Davis & Hedge 29854* (K, W). Çoruh gorge between Artvin and Ardanuç, rocky igneous banks, 450 m, 26 June 1957, *Davis & Hedge 30051B* (BM, E). C. 13 km s. of Borçka, on calcareous schist, c. 250 m, 10 July 1959, *Huber-Morath 15268* (Herb. Huber-Morath). Artvin hill above town, on dry banks in oak-juniper scrub, c. 1000 m, 1 July 1960, *Stainton & Henderson 5950* (E). Yusufeli, between Sarigöl and Barhal, rocky slopes of gorge, 900 – 950 m, 1 Aug. 1966, *Davis 47669* (E). Between Kars and Artvin, Kordevan Dagh, c. 1300 m, rock crevices, 3 Sept. 1966, *Furse 9144* (K).

U.S.S.R.: near Batum, *Calvert s.n.* (type). Between Keda and Chula, on dry rock, in wooded lower mountain regions, 21 June 1890, *Sommier & Levier 1064* (B, FI, F, WU). Batum near Artvin, 15 July 1904, *Nichailowsxy s.n.* (FI). Batum, Artvin, near Lomaschen, on steep slopes, 28 June 1907, *Woronow s.n.* (FI). Idem, *Woronow 273* (W, WU). Batum near Khula, 22 June – 5 July 1912, *Holmberg 1887* (W). Batum, Artvin, near Dsansul, c. 500 m, 5 – 18 July 1912, *Holmberg 2227* (K, W).

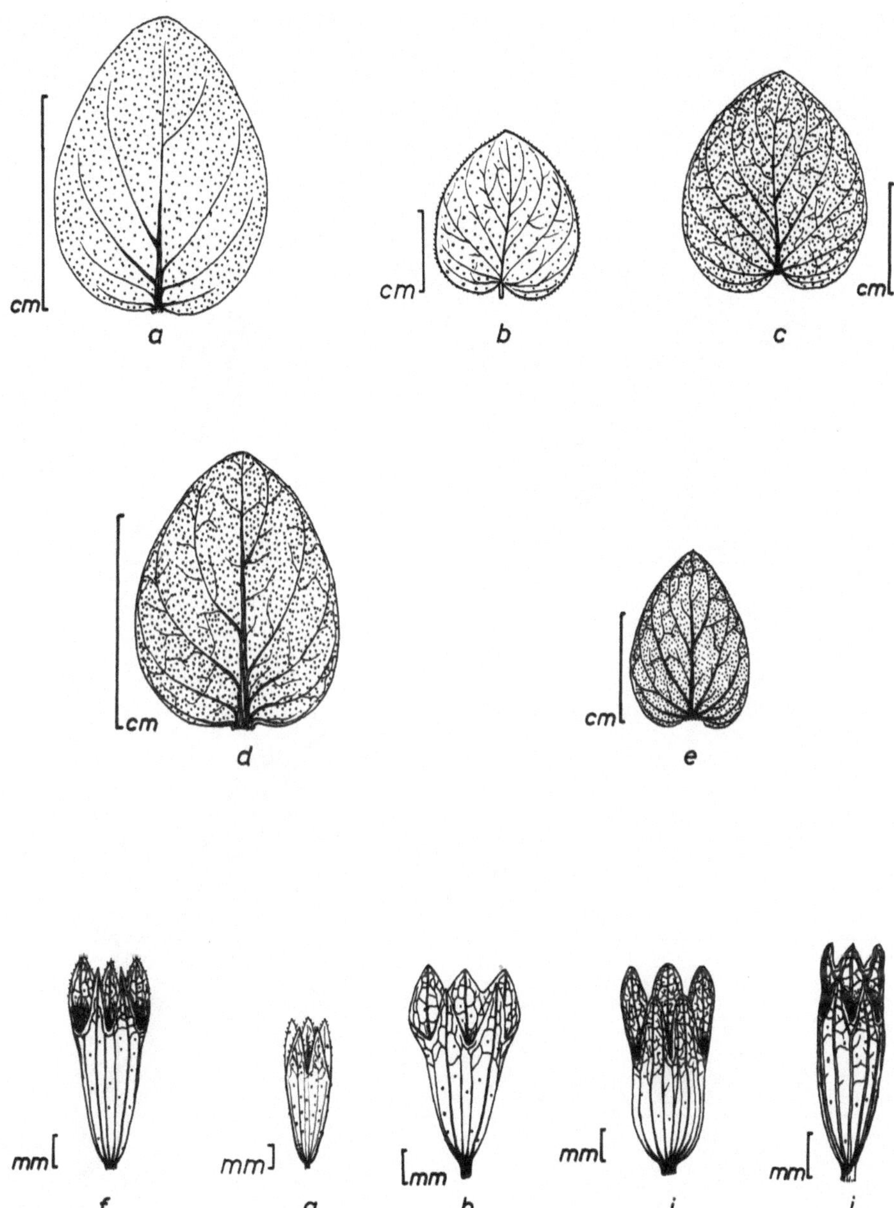

Figure 11. Leaves and calyces of the species in the section *Brevifilamentum*, except *O. rotundifolium* (for which see figure 14): a. and f. *O. acutidens;* b. and g. *O. bargyli;* c. and h. *O. brevidens;* d. and i. *O. haussknechtii;* e. and j. *O. leptocladum.*

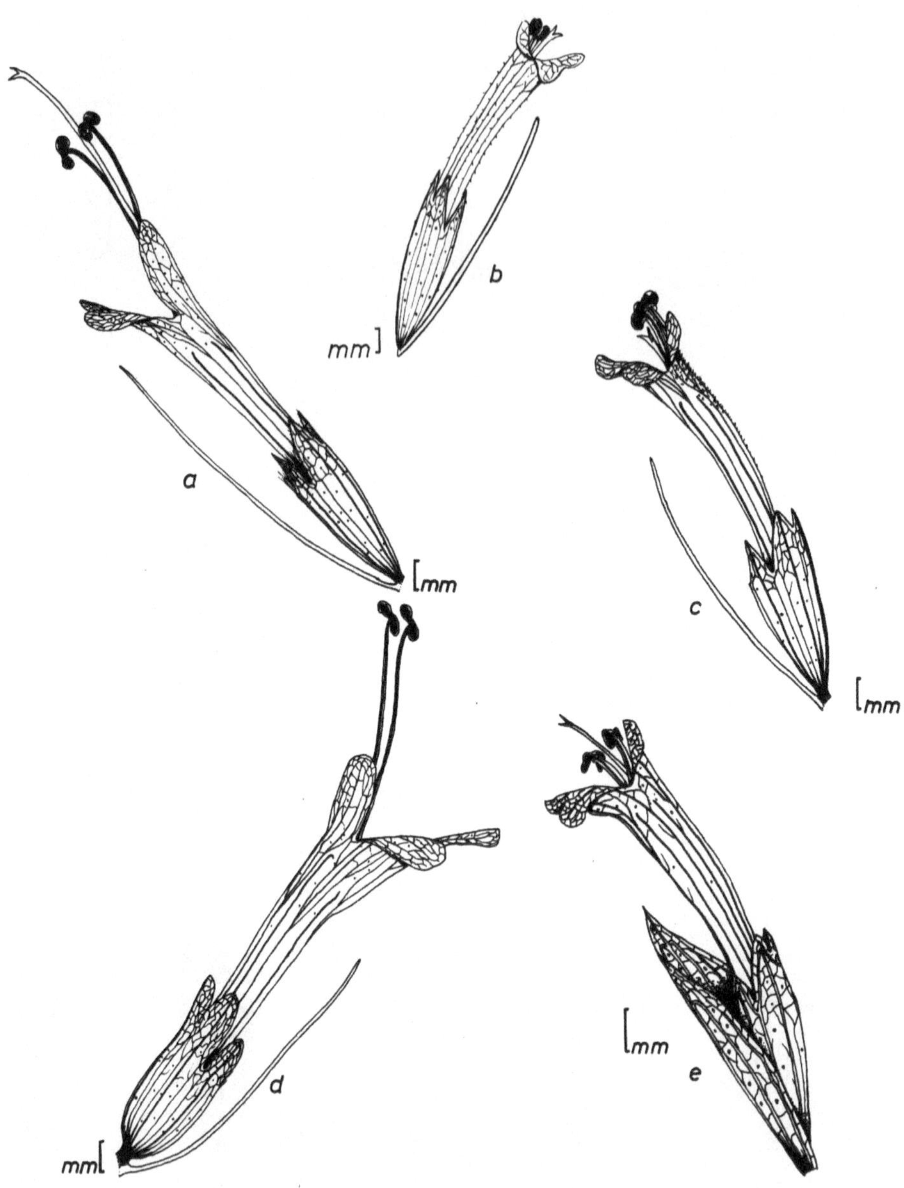

Figure 12. Flowers with bracts in side view of the species in the section *Brevifilamentum*, except *O. rotundifolium* (for which see figure 14): a. *O. acutidens;* b. *O. bargyli;* c. *O. brevidens;* d. *O. haussknechtii;* e. *O. leptocladum* (bract not cut).

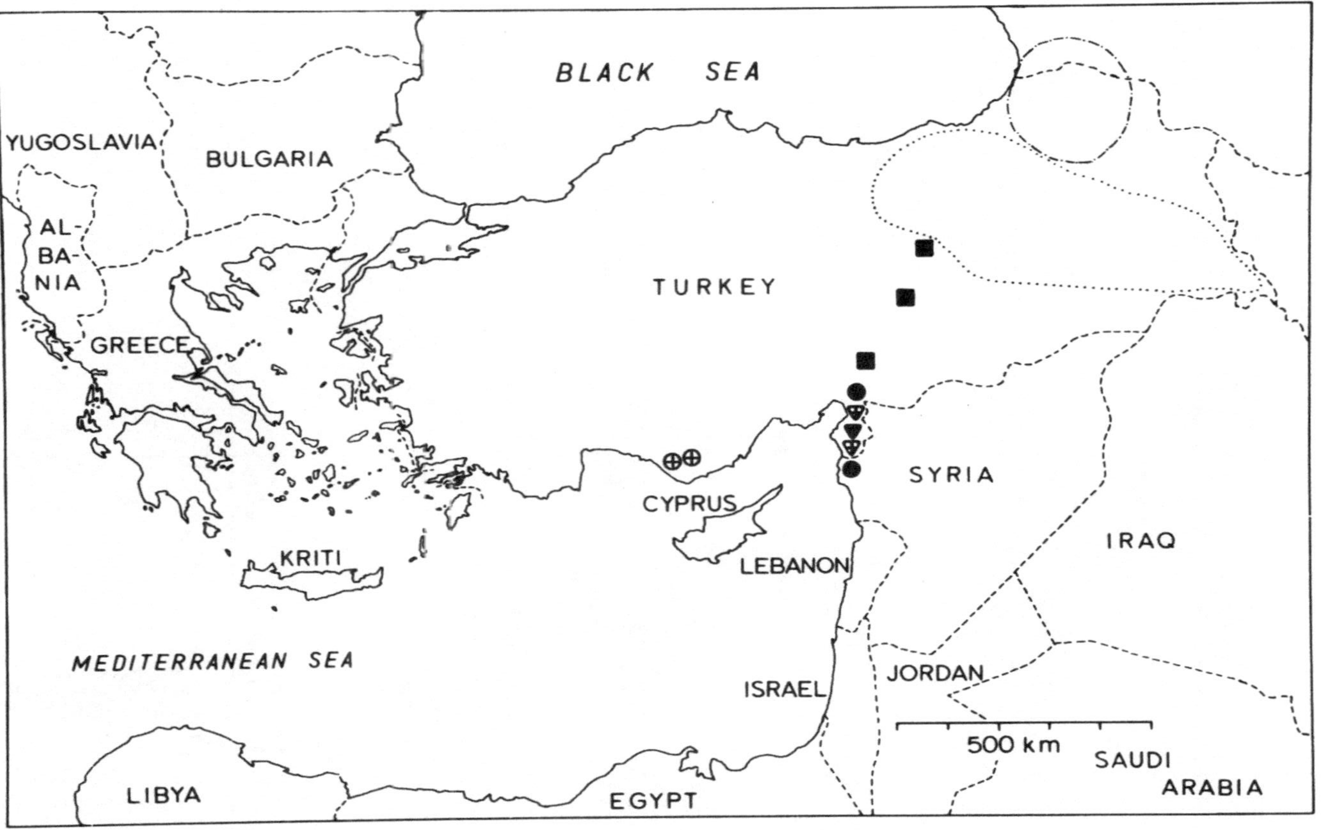

Figure 13. Distribution of the species in the sections *Brevifilamentum* and *Longitubus*: ...... *O. acutidens*; ● *O. bargyli*; ▼ *O. brevidens*; ■ *O. haussknechtii*; ⊕ *O. leptocladum*; _._._. *O. rotundifolium*; ⊽ *O. amanum*.

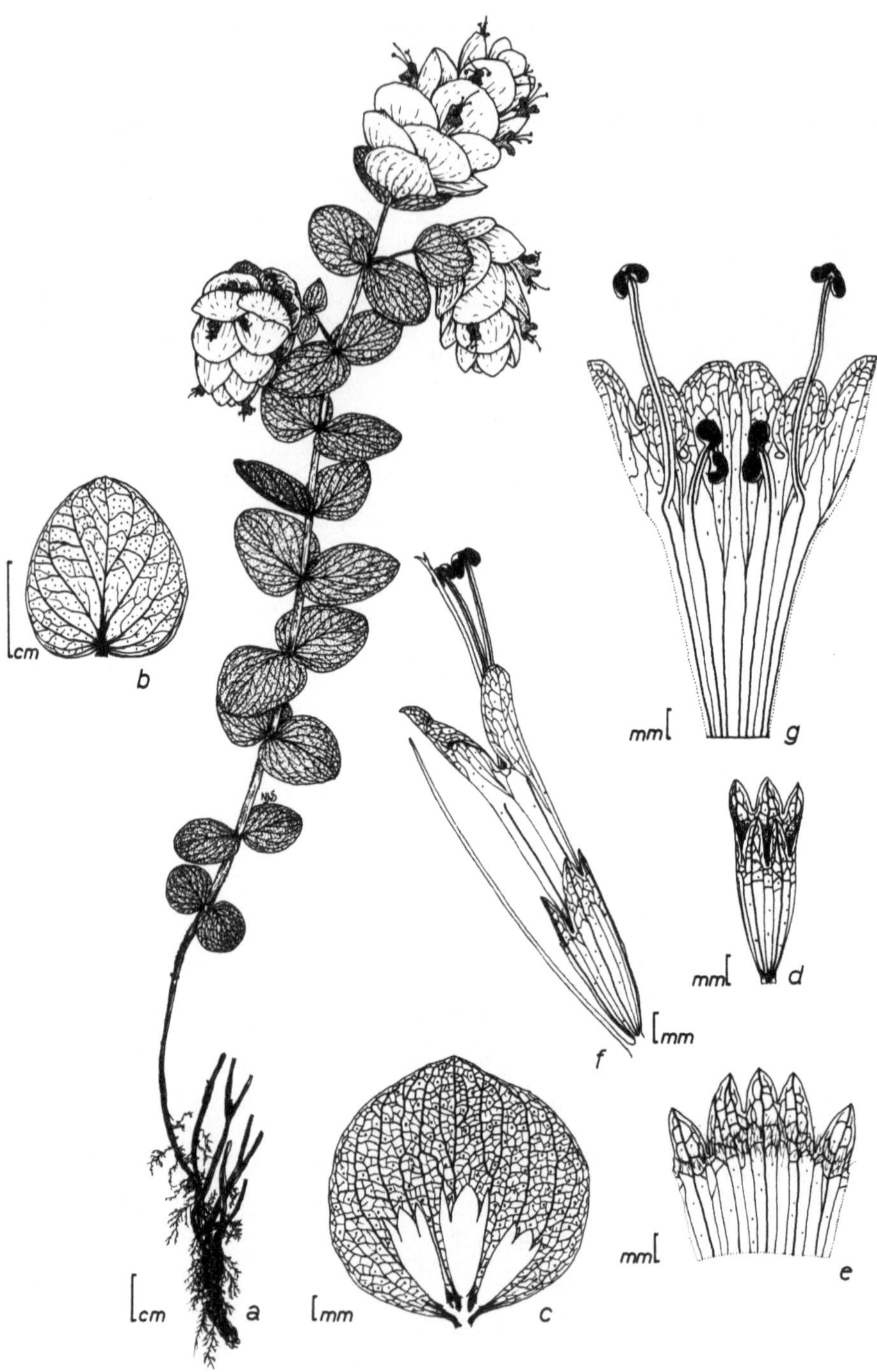

Figure 14. O. rotundifolium: a. habit; b. leaf; c. bract with calyces in upper lip view; d. calyx in lower lip view; e. calyx cut through the lower lip; f. flower with bract in side view; g. corolla cut through the lower lip.

IV. **Section Longitubus** Ietswaart

Section *Longitubus* Ietswaart, Not. R. B. G. Edinburgh 38(1): 47 (1980). – Type and only species: *Origanum amanum* Post.

Branches of the first order seldom present, those of the second order never so. *Leaves* more or less herbaceous. *Spikes* large, (nearly) not nodding. *Bracts* imbricate, c. 2 x calyces, membranous, slightly pilosellous. *Flowers* several per verticillaster and shortly pedicellate, bisexual, very large. *Calyces* tubular, 2-lipped for c. $\frac{2}{5}$, teeth in upper and lower lips well developed; throats pilose. *Corollas* 2-lipped for c. $\frac{1}{7}$, c. 4 x calyces, not saccate. *Stamens* nearly equal in length, included; filaments very short, c. $\frac{1}{50}$ × corollas.

20. Origanum amanum Post – **Figs. 13** and **15.**

O. amanum Post, Bull. Herb. Boiss. 3: 161 (1895); Post & Dinsmore, Fl. Syr. Palest. Sin. 2: 332 (1933); Thiebaut, Fl. Libano-Syr. 3: 46 (1953). *Amaracus amanus* (Post) Bornmüller, Notizbl. Bot. Gart. Berlin 7(63): 26 (1917); Jackson, in Hooker's Ic. Pl. 33: 1 (1933). – Type: *Post 323*, Turkey, Hatay, Gaiour Dagh (holo. G).

Subshrubs. Roots up to 1.5 cm in diameter. Young shoots hirsute or hirtellous. Stems often many together in a specimen, ascending and rooting at the bases, up to 20 cm long, light to purplish brown, hirsute to scabrous (hairs c. 0.8 mm long). Branches absent or a few present at the top of the stems, c. 1 cm long. *Leaves* up to 12 pairs per stem, (sub)sessile, heart-shaped, seldom ovate, tops acute or acuminate, 11 (6 – 19) mm long, 8 (4 – 14) mm wide, light green, slightly glaucous, hirsute to scabrous (hairs c. 0.5 mm long), sessile glands up to 500 per cm^2. *Spikes* subglobose, 25 (15 – 40) mm long, 18 (14 – 20) mm wide. *Bracts* 7 (3 – 9) pairs per spike, ovate or oval, seldom ± lanceolate, tops acuminate or acute, 14 (8 – 21) mm long, 9 (4 – 15) mm wide, vividly purple. *Flowers* 4 (2 – 10) per verticillaster, with c. 1 mm long pedicels. *Calyces* 8 (5 – 12) mm long, pilosellous outside and the lips also inside; upper lips divided, for $\frac{1}{10}$ – $\frac{4}{5}$, into 3 (sub)equal, deltoid to (narrowly) triangular, 1.0 (0.1 – 3.0) mm long teeth; lower lips slightly longer to slightly shorter than the upper lips, consisting of 2 (sub)equal, (narrowly) triangular, usually acute, 2.5 (1.0 – 3.5) mm long teeth. *Corollas* 30 (15 – 50) mm long, pink, outside densely pilosellous; tubes slightly curved downwards; upper lips divided, for $\frac{1}{10}$ – $\frac{3}{5}$, into 2, 1.4 (0.2 – 2.3) mm long lobes; lower lips divided, for c. $\frac{3}{5}$, into 3, somewhat unequal, roundish, 2.0 (0.1 – 3.5) mm long lobes. *Stamens* all 4 included; filaments c. 0.5 mm long. *Styles* only slightly protruding, up to 42 mm long. Chromosome number 2n = 30.

Geography and ecology. *O. amanum* is found in a few places in the Amanus Mts., where it occurs from 1500 – 2200 m, on calcareous rocks and slopes. It flowers from June to September.

Notes. 1. For its small area of distribution, *O. amanum* shows a large variation in the shape and size of the bracts, calyces and corollas. 2. In some specimens a very large (c. 98) percentage of abortive pollen was found, and in some others a nearly regular 5-toothed calyx. Both observations point to a recent hybridization in some of the populations. 3. From *O. bargyli* and *O. brevidens*, *O. amanum* differs in its much longer corollas with four subsessile stamens. 4. Once a hybrid has been found between *O. amanum* and *O. laevigatum* (sect. *Prolaticorolla*) (see p. 135). An artificial hybrid has been made from *O. amanum* and *O. dictamnus* (see p. 140).

TURKEY. PROV. HATAY: Gaiour-Dagh (Amanus), c. 1500 m, Aug. – Sept. 1892, *Post 323* (type). Dildil Dagh, 1700 – 2300 m, July 1908, *Haradjian 2433* (G). Dildil Dagh, 1500 – 2000 m, Aug. 1911, *Haradjian 3884* (G, E, W). Dildil Dağ between Başkonuş Y. and Huseyin Oluk Çe., on rocks in steep gulley, 1800 m, 27 Aug. 1949, *Davis 16390* (E, K). Dildil Dağ, above Atlik Y., on sloping limestone rocks, 2000 m, 27 Aug. 1949, *Davis 16439* (W).

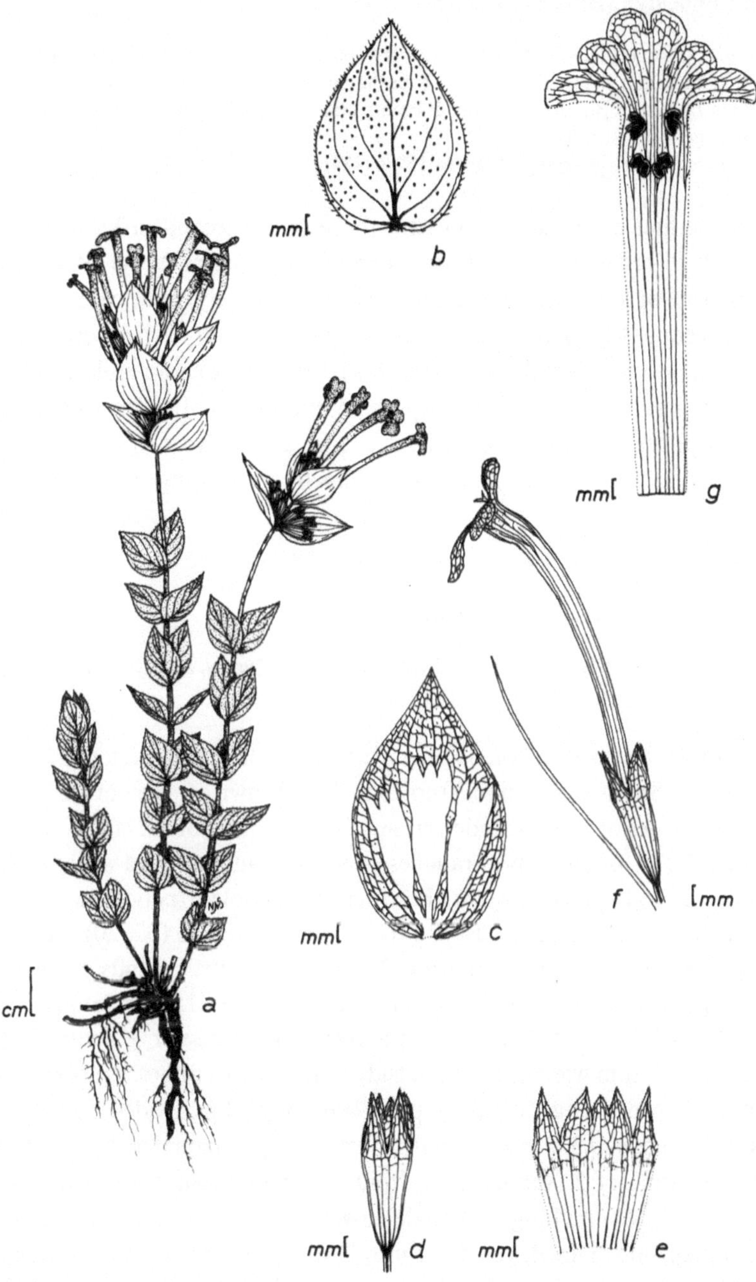

Figure 15. O. amanum: a. habit; b. leaf; c. bract with calyces in upper lip view; d. calyx in lower lip view; e. calyx cut through the lower lip; f. corolla with bract in side view; g. corolla cut through the lower lip.

V. **Section Chilocalyx** (Briquet) Ietswaart

Section *Chilocalyx* (Briquet) Ietswaart *comb. nov.* – *Majorana* section *Chilocalyx* Briquet, in Engler & Prantl, Nat. Pflanzenfam. 4(3a): 307 (1895). – Type designated here: *Origanum micranthum* Vogel.
Majorana section *Holocalyx* Briquet, in Engler & Prantl, Nat. Pflanzenfam. 4(3a): 307 (1895). – Type: *Origanum microphyllum* (Bentham) Vogel.
Subgenus *Majorana* (Miller) Vogel p.p., Linnaea 15: 76 (1841).

Branches of the first order always present, those of the second order usually so, those of the third order often so. *Leaves* herbaceous. *Spikes* (very) small, erect. *Bracts* often more or less continuing from the leaves, (densely) imbricate, c. $\frac{4}{5}$ – $1\frac{1}{5}$ x calyces, herbaceous, whitish, greyish or green, hairy. *Flowers* 2 per verticillaster, bisexual or female, (very) small. *Calyces* tubular or slightly campanulate, 1- or 2-lipped for $\frac{1}{5}$ – $\frac{2}{5}$; small teeth or lobes in upper and lower lips often present; throats conspicuously pilose. *Corollas* 2-lipped for c. $\frac{2}{5}$, c. 2 × calyces. *Stamens* slightly unequal in length, straight, (sub)included or shortly protruding; filaments c. $\frac{2}{5}$ × corollas.

21. **Origanum bilgeri** Davis – **Figs. 16** and **17.**

O. bilgeri Davis, Kew Bull. 1949: 406 (1949). – Type: *Davis 14720*, Turkey, Antalya, near Geyik Dağ (holo. K, iso. E, para. G, JE, W).

Subshrubs, flowers bisexual or female only. Roots up to 0.5 cm in diameter. Young shoots tomentose. Stems erect, up to 30 cm long, light brown, ± tomentose (hairs c. 1 mm long). Branches of the first order present in the upper $\frac{2}{5}$ of the stems, up to 10 pairs per stem, 2 (1 – 6) cm long; branches of the second and third order usually present. *Leaves* up to 18 pairs per stem, the lower ones petiolate (petioles up to 6 mm long), roundish or ovate, tops obtuse, 14 (6 – 23) mm long, 13 (5 – 20) mm wide, greyish, ± pubescent or tomentose (hairs c. 0.8 mm long), sessile glands inconspicuous, up to 650 per cm^2. *Spikes* subglobose or cylindrical, 6 (3 – 15) mm long, c. 4 mm wide. *Bracts* 6 (3 – 12) pairs per spike, ovate to oval, tops obtuse or ± acute, 2.5 (2 – 4) mm long, 2 (1 – 3) mm wide, greyish, outside tomentose or pubescent. *Calyces* 2-lipped for c. $\frac{1}{5}$, c. 2.5 mm long, outside ± pubescent; upper lips divided, for c. $\frac{1}{2}$, into 3 (sub)equal, deltoid, c. 0.3 mm long teeth; lower lips approximately as long as the upper lips, consisting of 2 (sub)equal, deltoid, c. 0.5 mm long teeth. *Corollas* 2-lipped for c. $\frac{2}{5}$, 5 (3 – 6) mm long (in female flowers c. 3.5 mm long), white, outside pilosellous; upper lips divided, for c. $\frac{1}{10}$, into 2, c. 0.2 mm long lobes; lower lips divided, for c. $\frac{3}{5}$, into 3 subequal, c. 1 mm long lobes. *Stamens* (sub)included; filaments up to 1.5 and 2 mm long. *Styles* up to 8 mm long.

Geography and ecology. *O. bilgeri* is only known from the type collection, so little can be said about its ecology etc.
Note. *O. bilgeri* is closely related to *O. minutiflorum*, from which it differs in its more or less tomentose indumentum and its somewhat larger leaves, bracts and flowers.

When more specimens of both species become available from different sites, it is possible that they can better be united into one species. For the moment they are best treated as different species. *O. bilgeri* is also related to *O. micranthum*, from which it differs in its larger, less tomentose leaves and its white, somewhat larger corollas.

TURKEY. PROV. ANTALYA: Han Boğaz forest near Geyik Dağ, in open Cedretum on s. side of dry river-bed, c. 1500 m, 30 Aug. 1947, *Davis 14720* (type).

22. **Origanum micranthum** Vogel – **Figs. 17** and **18.**

O. micranthum Vogel, Linnaea 15: 77 (1841); Boissier, Fl. Or. 4: 552 (1879). *Majorana micrantha* (Vogel) Briquet, in Engler & Prantl, Nat. Pflanzenfam. 4(3a): 307 (1895). Type: *Kotschy 471*, Turkey, Adana, Taurus Mts. (holo. W).

Subshrubs, flowers bisexual or female only. Roots up to 1.5 cm in diameter. Young shoots tomentose. Stems ascending and rooting at the bases or erect, up to 35 cm long, light brown, tomentose (hairs c. 1 mm long). Branches of the first order present in the upper $\frac{1}{2}$ of the stems, up to 10 pairs per stem, 3.5 (1.5 – 13) cm long; branches of the second order usually present, those of the third order sometimes so. *Leaves* up to 18 pairs per stem, the lower ones petiolate (petioles up to 5 mm long), roundish or ovate to oval, tops $\pm$ obtuse, 8 (3 – 14) mm long, 6 (2 – 12) mm wide, whitish, tomentose (hairs c. 0.8 mm long), sessile glands inconspicuous, up to 250 per cm^2. *Spikes* $\pm$ ovoid (seldom cylindrical), 4 (3 – 10) mm long, c. 3 mm wide. *Bracts* 4 (2 – 9) pairs per spike, ovate to obovate, tops $\pm$ obtuse, 3 (1.5 – 4) mm long, 2.5 (1.5 – 3) mm wide, whitish, outside tomentose. *Calyces* 2-lipped for $\frac{1}{5}$ – $\frac{2}{5}$, c. 2 mm long, outside tomentose; upper lips divided, for c. $\frac{2}{5}$, into 3 (sub)equal, $\pm$ deltoid, c. 0.2 mm long teeth or lobes, or (sub)entire; lower lips approximately as long as the upper lips, consisting of 2 (sub)equal, deltoid, 0.4 (0.2 – 0.5) mm long teeth. *Corollas* 2-lipped for c. $\frac{2}{5}$, 3.5 (2.5 – 5) mm long (in female flowers c. 3 mm long), purple or pink, outside pilosellous; upper lips divided, for c. $\frac{1}{5}$, into 2, c. 0.3 mm long lobes; lower lips divided, for c. $\frac{3}{5}$, into 3 subequal, c. 1 mm long lobes. *Stamens* (sub)included; filaments up to 1.5 and 2 mm long. *Styles* up to 7 mm long.

Geography and ecology. *O. micranthum* is known from one place only in southern Turkey: the environs of the Cilician Gates, where it grows at an altitude of c. 1500 m and flowers from July to September.

Notes. 1. *O. micranthum* is related to *O. bilgeri* and *O. minutiflorum*. From the first species it differs in its smaller leaves, more tomentose indumentum, smaller spikes and purple or pink corollas. From the latter species it differs in its tomentose indumentum and purple or pink corollas, and also in its larger bracts. 2. A putative hybrid, from nature, has been found between *O. micranthum* and *O. vulgare* ssp. *hirtum* (see p. 141).

TURKEY. PROV. ADANA: Taurus Mts., summer 1836, *Kotschy 471* (type). Environs of the castle near Güllek, at c. 1500 m, 23 Aug. 1853, *Kotschy 261* (G, JE, W). Cilician Gates, on rocks near the ruins of the castle, Aug. 1855, *Balansa 537* (p.p.) (G, JE). Güllek Gala, c. 1400 m, *Siehe 669* (BM, E, G, JE).

23. **Origanum microphyllum** (Bentham) Vogel – **Figs. 16** and **17.**

O. microphyllum (Bentham) Vogel, Linnaea 15: 76 (1841); Boissier, Fl. Or. 4: 552 (1879); Tutin et al., Fl. Eur. 3: 172 (1972). *Majorana microphylla* Bentham, Lab. Gen. Sp.: 338 (1834). Type: *Sieber s.n.*, Greece, Kriti (holo. W, iso. E, L, W).

O. maru sensu Sibthorp et Smith (non Linnaeus), Fl. Graeca 6: 59 (1826); Halácsy, Consp. Fl. Graec. 2: 556 (1902); *Majorana maru* (Sibthorp et Smith) Hayek, Prodr. Fl. Penins. Balc. 2: 336 (1931); Rechinger, Fl. Aegaea: 553 (1943).

Subshrubs, flowers bisexual. Roots up to 1.5 cm in diameter. Young shoots tomentellous. Stems usually ascending, ramified and rooting at the bases, slender, up to 45 cm long, light or purplish brown, glabrous, or sparsely tomentellous (hairs c. 0.4 mm long). Branches of the first order present in the upper $\frac{1}{2}-\frac{4}{5}$ of the stems, up to 14 pairs per stem, 2 (0.5 – 11) cm long, only the upper ones bearing spikes; branches of the second order sometimes present. *Leaves* up to 20 pairs per stem (the lower ones often soon falling), shortly petiolate (petioles up to 4 mm long), roundish or ± heart-shaped, or ovate, tops usually obtuse, 5 (2 – 13) mm long, 4 (2 – 11) mm wide, whitish, tomentellous (hairs c. 0.2 mm long), sessile glands inconspicuous, up to 300 per cm^2. *Spikes* often 3 closely together at a branch, subglobose, ovoid or cylindrical, 7 (4 – 14) mm long, c. 4 mm wide. *Bracts* 6 (2 – 12) pairs per spike, ovate to obovate, tops obtuse, 3.5 (2 – 5.5) mm long, 2.5 (2 – 4.5) mm wide, whitish, outside tomentellous. *Calyces* usually 1-lipped for c. $\frac{1}{5}$, 2.5 (1.5 – 3) mm long, outside ± glabrous; upper lips (sub)entire; lower lips sometimes consisting of 2 very small lobes. *Corollas* 2-lipped for c. $\frac{2}{5}$, 5 (3 – 7.5) mm long, purple; upper lips divided, for c. $\frac{1}{10}$, into 2, c. 0.2 mm long lobes; lower lips divided, for c. $\frac{3}{5}-\frac{4}{5}$, into 3 subequal, 1.2 (0.6 – 2.0) mm long lobes. *Stamens* shortly protruding; filaments up to 4 and 5 mm long. *Styles* up to 9 mm long.

Geography and ecology. *O. microphyllum* occurs in several places on Kriti, where it grows on rocky soils and in crevices, on limestone, from 500 – 1700 m. It flowers from June to September.

Notes. 1. *O. microphyllum* holds a distinct position in the section *Chilocalyx*, owing to its slender stems, small leaves and usually entire calyces. It is not closely related to the other three species in the section. 2. Sieber introduced the name *O. microphyllum* on herbarium labels. 3. The specimen of *O. maru* in the Linnaean herbarium (no. 743.12) should be reckoned to *O. syriacum*. Authors who described the taxon *O. microphyllum* under the name *O. maru* sometimes have been cited with this *O. maru* "non L." (e.g. Sibthorp & Smith and Hayek). 4. Occasionally *O. microphyllum* hybridizes with *O. vulgare* ssp. *hirtum* (see p. 139).

GREECE. KRITI: Mt. Sphak near Omalo, on rocks, c. 1200 m, June 1846, *Heldreich s.n.* (COI, G). Mt. Lakus, 5 July 1883, *Reverchon 128* (G, JE, W). Khania, above Omaló, on calcareous rocks, 1893,

Baldacci 85 (BM, W). Pedeada, on rocks, 1899, *Baldacci 91* (BM, W). Lassithi, on rocks, 1 July 1899, *Baldacci 345* (BM). Mt. Lassithi, on rocks, 1 July 1904, *Leonis & Halácsy 230* (COI, W). Lassithi, environs of the upland plain, c. 1000 m., on rocky soils, 23 July 1904, *Dörfler 1086* (G, JE). Mt. Lakus, June 1932, *Guiol 2119* (BM). Lassithi, 600 – 1500 m, 19 June 1937, *Lemperg 511* (W). Lassithi, 18 July 1939, *Regel s.n.* (G). Near Omalo, 11 Aug. 1939, *Regel s.n.* (G). Sphakia, Levka Ori, between Samaria and hag. Nikolaos, in stony places, c. 500 m, 4 June 1942, *Rechinger 13741* (G, BM). Selinos, Levka Ori, chain of Xyloskala, on calcareous rocks, c. 1250 m, 12 June 1942, *Rechinger 13685* (BM, G). Lassithi, near Messa Lassithi, on calcareous rocks, c. 900 m, 11 July 1942, *Rechinger 14373* (BM, G). Levka Ori, in the higher regions, *Pinatzi 14024* (G). Lassithi, alpine pasture Meriussa, above chapel Aj. Pelajia, on limestone soils, c. 1200 m, 26 June 1961, *Greuter S3709* (G). Levka Ori, Samarian Gorge, 7 June 1966, *Whitefoord 119* (BM).

24. **Origanum minutiflorum** Schwarz et Davis – **Figs. 16** and **17.**

O. minutiflorum Schwarz et Davis, Kew Bull. 1949: 408 (1949). – Type: *Davis 14185*, Turkey, Antalya, Tahtali Dağ (holo. K, iso. E, para. G, W).

Subshrubs, flowers bisexual or female only. Roots up to 0.5 cm in diameter. Young shoots ± tomentose. Stems usually erect, up to 35 cm long, light brown, pilose (hairs curved and appressed, c. 0.8 mm long). Branches of the first order present in the upper $\frac{2}{5}$ of the stems, up to 10 pairs per stem, 2 (0.5 – 4) cm long; branches of the second order usually present, those of the third order sometimes so. *Leaves* up to 18 pairs per stem, shortly petiolate (petioles up to 6 mm long), ovate or oval, tops ± acute or obtuse, 10 (3 – 16) mm long, 5 (2 – 12) mm wide, green, pilosellous (hairs curved and appressed, c. 0.4 mm long), sessile glands up to 1500 per cm^2. *Spikes* subglobose to cylindrical, sometimes rather loose at the bases, 4 (2 – 8) mm long, c. 3 mm wide. *Bracts* 4 (3 – 6) pairs per spike, ovate or oval, 2 (1 – 3) mm long, 1 (0.5 – 1.5) mm wide, green, outside pilosellous. *Calyces* 2-lipped for c. $\frac{1}{5}$, c. 2 mm long, outside pilosellous; upper lips divided, for c. $\frac{2}{5}$, into 3 (sub)equal, ± deltoid, c. 0.2 mm long lobes; lower lips consisting of 2 (sub)equal, deltoid, c. 0.4 mm long teeth. *Corollas* 2-lipped for c. $\frac{2}{5}$, 3 (2 – 4) mm long (in female flowers c. 2.5 mm long), white; upper lips divided, for c. $\frac{1}{5}$, into 2, c. 0.2 mm long lobes; lower lips divided, for c. $\frac{4}{5}$, into 3 subequal, c. 1 mm long lobes. *Stamens* subincluded; filaments up to 1.5 and 2 mm long. *Styles* up to 6 mm long.

Geography and ecology. *O. minutiflorum* is known from a few places in southern Turkey, where it grows on rocky slopes on limestone, at a height of c. 1600 m. It has been found flowering in July and August.

Notes. 1. As mentioned, *O. minutiflorum* is closely related to *O. bilgeri*, from which it differs in its smaller leaves, bracts and corollas, and also in its not tomentose, green, more glandular punctate leaves. A few specimens have been collected (labelled *O. minutiflorum*), which are nearly intermediate between the two species (see also note for *O. bilgeri*). 2. From the also related *O. micranthum*, *O. minutiflorum* differs in its green, pilosellous, more glandular punctate leaves, and white corollas.

TURKEY. PROV. ANTALYA: Tahtali Dağ near Çukur Yayla, rocky limestone slopes, c. 1500 m, 15 Aug. 1947, *Davis 14185* (type). Bozburun Dağ, at Taşli Yayla, open rocky slopes, c. 1700 m, 27 July 1949, *Davis 15773* (E). Distr. Elmali, Koçeva, July 1964, *Demirdöğen 2580* (E).

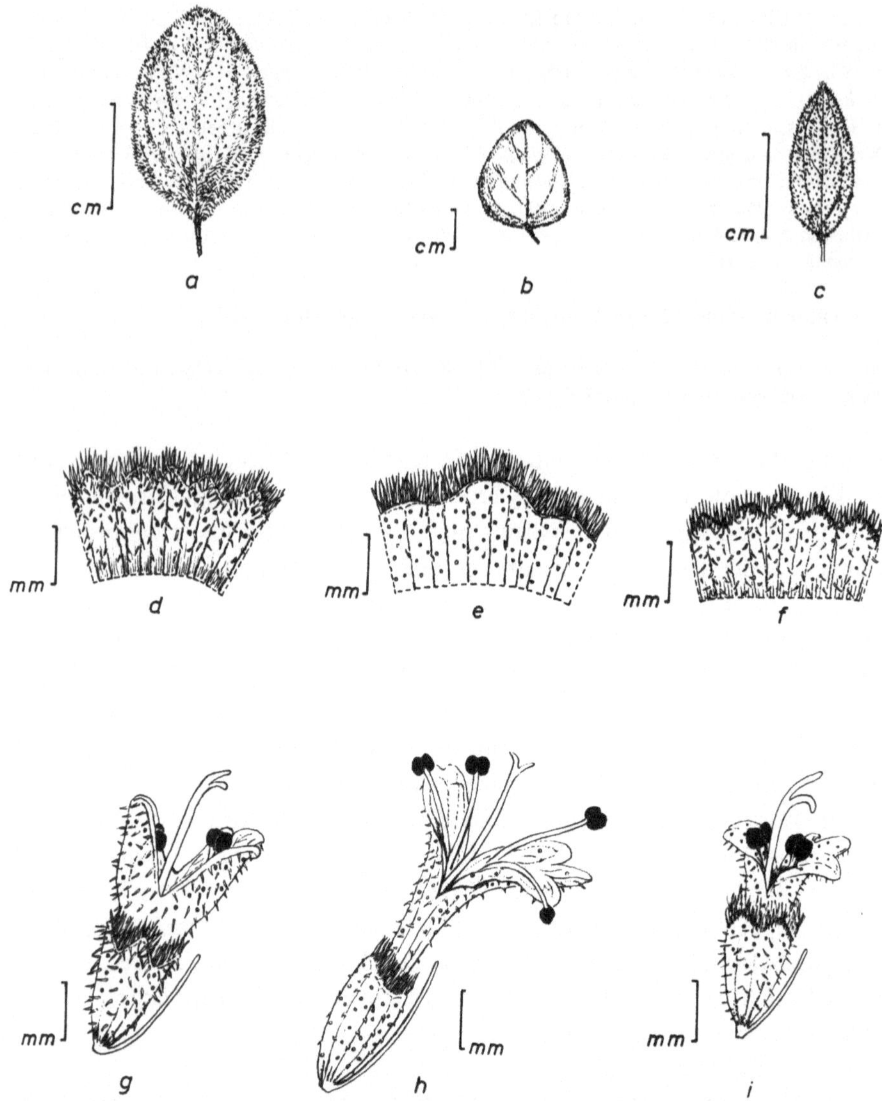

Figure 16. Leaves, calyces cut through the lower lip, and flowers with bracts in side view of the species in the section *Chilocalyx*, except *O. micranthum* (for which see figure 18): a., d. and g. *O. bilgeri;* b., e. and h. *O. microphyllum;* c., f. and i. *O. minutiflorum*.

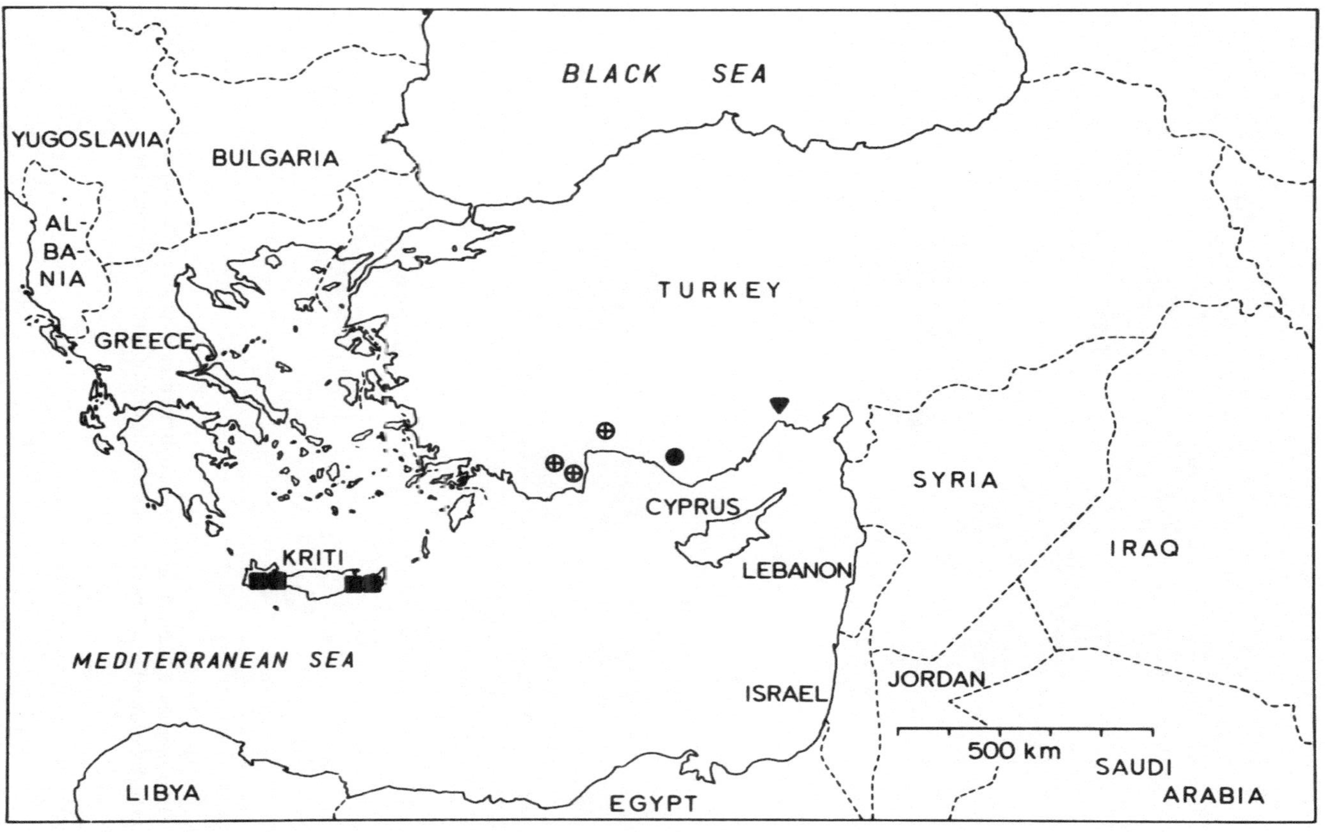

Figure 17. Distribution of the species in the section *Chilocalyx*: ● *O. bilgeri*; ▼ *O. micranthum*; ■ *O. microphyllum*; ⊕ *O. minutiflorum*.

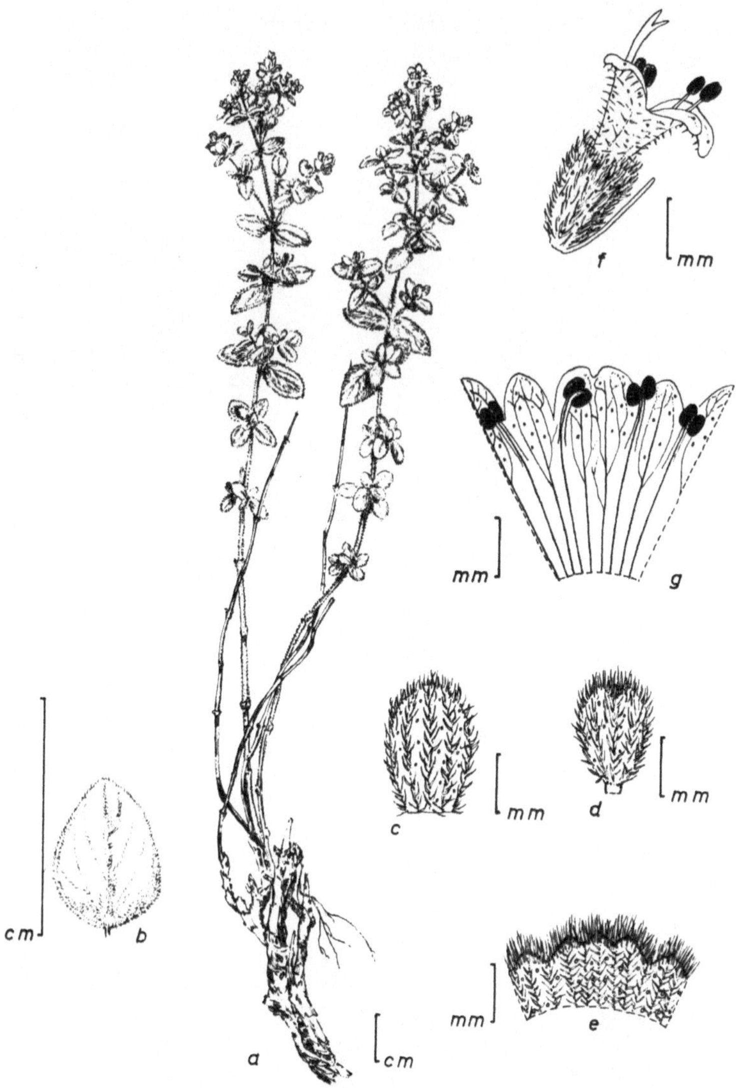

Figure 18. O. micranthum: a. habit; b. leaf; c. bract outside; d. calyx in lower lip view; e. calyx cut through the lower lip; f. flower with bract in side view; g. corolla cut through the lower lip.

VI. Section Majorana (Miller) Bentham

Section *Majorana* (Miller) Bentham, in de Candolle, Prodr. Syst. Nat. 12: 195 (1848). – Type designated here: *Origanum majorana* Linnaeus.
Majorana section *Schizocalyx* (Scheele) Briquet, in Engler & Prantl, Nat. Pflanzenfam. 4(3a): 307 (1895).
Subgenus *Majorana* (Miller) Vogel p.p., Linnaea 15: 76 (1841).

Branches of the first order always present, those of the second order often so, and those of the third order sometimes so. *Leaves* herbaceous. *Spikes* (sub)globose, often quadrigonus-cylindrical, small, erect. *Bracts* different from the leaves, densely imbricate, ± as long as calyces, enclosing these marginally, herbaceous, whitish, greyish or green, hairy. *Flowers* 2 per verticillaster, bisexual or female, small. *Calyces* flattened, 1-lipped for $\frac{9}{10}$ or more; upper lips entire or denticulate; throats not pilose. *Corollas* usually 2-lipped for c. $\frac{2}{5}$, c $2\frac{1}{2}$ × calyces, flattened. *Stamens* unequal in length, divergent or straight, (shortly) protruding; filaments c. $\frac{3}{5}$ × corollas.

25. Origanum majorana Linnaeus – Figs. 19 and 20.

O. majorana Linnaeus, Sp. Pl.: 590 (1753); Aiton, Hort. Kew. 2: 312 (1789); Desfontaines. Fl. Atlantica 2: 27 (1799); Röhling & Koch, Deutschl. Fl. 4: 306 (1833); Fraas, Syn. Pl. Fl. Class.: 182 (1845); Willkomm & Lange, Prodr. Fl. Hisp. 2: 399 (1868); Boissier, Fl. Or. 4: 553 (1879); Nyman, Consp. Fl. Eur.: 592 (1881); Hooker, Fl. Brit. India 4: 648 (1885); Battandier & Trabut, Fl. Alg. 2: 675 (1884); Wittmack, Verh. Bot. Ver. Brandenburg 32: 24 (1890); Wittmack, Verh. Bot. Ver. Brandenburg 33: 44 (1891); Beck von Mannagetta, Fl. Nieder-Österr. 2: 993 (1893); Bonnet & Barrate, Cat. Pl. Vasc. Tunesie: 328 (1896); Halácsy, Consp. Fl. Graec. 2: 557 (1902); Holmes, Perf. Ess. Oil Rec. 4: 69 (1913); Appl, Preslia 6: 3 (1928); Post & Dinsmore, Fl. Syr. Palest. Sin. 2: 335 (1933); Fiori, Nuova Fl. Anal. It. 2: 456 (1969). Tutin et al., Fl. Eur. 3: 171 (1972). *Thymus majorana* (Linnaeus) Kuntze, Taschenfl. Leipzig: 106 (1867). *Majorana majorana* (Linnaeus) Karsten, Deutsche Fl.: 999 (1882). *Amaracus majorana* (Linnaeus) Schinz et Thellung, Bull. Herb. Boiss. 7: 576 (1907). – Type: *Linnaeus s.n.* (holo. BM).
Majorana vulgaris Miller, Gard. Dict. Abr. IV Ed.: 829 (1754).
Amaracus vulgaris Hill, Brit. Herb.: 381 (1756).
Majorana hortensis Moench. Meth. Pl.: 406 (1794); Walpers, Rep. Bot. Syst. 3: 696 (1844); Briquet, in Engler & Prantl, Nat. Pflanzenfam. 4(3a): 307 (1895); Gams, in Hegi Ill. Fl. Mittel-Eur. 5: 2334 (1927); Hayek, Prodr. Fl. Penins. Balc. 2: 335 (1931); Jahandiez & Maire, Cat. Pl. Maroc 3: 650 (1943); Chevallier, Rev. Bot. Appl. 18: 593 (1938); Coutinho, Fl. Port.: 612 (1939); Wolf, Baileya 2: 64 (1954); Briquet, Prodr. Fl. Corse 3: 216 (1955).
O. odorum Salisbury, Prodr. Stirp.: 85 (1796).
O. majoranoides Willdenow, Sp. Pl. 3: 137 (1800). *O. majorana* L. var. *majoranoides* (Willdenow) Wittmack, Verh. Bot. Ver. Brandenburg 32: 28 (1890). – Type: *Willdenow s.n.* (holo. B).
Majorana ovalifolia Stokes, Bot. Mat. Med. 3: 350 (1812).
Majorana ovatifolia Stokes, Bot. Mat. Med. 3: 352 (1812).
Majorana tenuifolia Gray, Nat. Arr. Brit. Pl. 2: 381 (1821).
O. majorana Linnaeus var. *fruticulosa* Reichenbach, Fl. Germ. Exc.: 313 (1831).
Majorana cretica Kosteletzky, Med.-Pharm. Fl. 3: 769 (1834).
Majorana fragrans Rafinesque, Fl. Tell. 3: 86 (1836).
Majorana suffruticosa Rafinesque, Fl. Tell. 3: 86 (1836).
O. confertum Savi, Oss. Gen. Orig.: 12 (1840).
O. majorana Linnaeus var. *obovatum* Rafinesque, Aut. Bot.: 119 (1840).
O. suffruticosum hort. ex Steudel, Nomencl. Bot. 2: 227 (1841).
Majorana mexicana Martius, Bull. Acad. Roy. Brux. 11: 191 (1844).

O. dubium Boissier, Fl. Or. 4: 553 (1879); Halácsy, Consp. Fl. Graec. 2: 556 (1902). *Majorana dubia* (Boissier) Briquet, in Engler & Prantl, Nat. Pflanzenfam. 4(3a): 307 (1895); Hayek, Prodr. Fl. Penins. Balc. 2: 336 (1931); Rechinger, Fl. Aegaea: 533 (1943). *O. syriacum* Linnaeus ssp. *dubium* (Boissier) Holmboe, Stud. Veg. Cyprus: 162 (1914). – Type: *Lenormand s.n.*, Greece, Naxos (holo. G).

Subshrubs. Roots up to 1 cm in diameter. Young shoots tomentellous. Stems usually erect or ascending, sometimes ramified at the bases, up to 80 cm long, light or dark brown, usually tomentellous (hairs c. 0,4 mm long). Branches of the first order present in the upper $\frac{1}{5}-\frac{2}{5}$ of the stems, up to 10 pairs per stem, 1.5 (0.3 – 14) cm long; branches of the second order usually present, those of the third order sometimes so. *Leaves* up to 30 pairs per stem, more or less petiolate (petioles up to 15 mm long), roundish to ovate or oval, tops usually obtuse, 13 (3 – 35) mm long, 11 (2 – 30) mm wide, whitish or greyish, ± tomentellous (hairs c. 0.2 mm long), sessile glands inconspicuous, up to 1500 per cm^2; veins usually inconspicuous and not raised at the undersides. *Spikes* often 3 or 5 closely together at a branch, (sub)globose, ovoid or quadrigonus-cylindrical, 6 (3 – 20) mm long, c. 3 mm wide. *Bracts* 6 (2 – 30) pairs per spike, oval, obovate or ± rhomboid, tops usually obtuse and entire (sometimes acute or denticulate), 3 (2 – 4) mm long, 2 (1 – 3) mm wide, whitish or greyish, outside tomentellous; margins enclosing the calyces at the bases. *Calyces* 1-lipped for c. $\frac{9}{10}$, ± rhomboid, 2.5 (2 – 3.5) mm long, outside ± tomentellous; upper lips usually entire (sometimes denticulate). *Corollas* 2-lipped for c. $\frac{2}{5}-\frac{1}{2}$, 5 (3 – 7) mm long (in female flowers c. 3.5 mm long), white, when dried often yellowish, outside pilosellous; upper lips divided, for c. $\frac{1}{10}-\frac{1}{5}$, into 2, c. 0.2 mm long lobes; lower lips divided, for c. $\frac{4}{5}$, into 3 subequal, 1.3 (0.5 – 2.0) mm long lobes. *Stamens* (shortly) protruding; filaments up to 4 and 5 mm long. *Styles* up to 9 mm long.

Geography and ecology. *O. majorana* is native on Cyprus and the adjacent part of southern Turkey. Subspontaneously it also occurs in several other Mediterranean countries, e.g.: Yugoslavia, Italy, Corsica, southern Spain and Portugal, Morocco and Algeria. It is also cultivated in many countries in Europe, America and Asia, where it sometimes escapes from the gardens and is found subspontaneously. In its native and subspontaneous habitats it usually grows in dry, rocky (limestone) places, from 100 – 1500 m. It flowers from May to September.

Notes. 1. There is no doubt that *O. dubium* and *O. majoranoides* are names for specimens of *O. majorana* from native sites. 2. The description as given above is based on specimens from natural and subspontaneous populations throughout the Mediterranean area. 3. *O. majorana* has been cultivated for at least two centuries in gardens in western Europe, outside or in pots, as a (medicinal) herb. These cultivated specimens differ to some extent from natural ones, due to direct climatic influences, and also a certain degree of selection. Sometimes they behave as annual or biennial herbs with a less compact habitus, longer branches and a less dense indumentum, whilst the leaves are larger and longer petiolate. 4. In *O. majorana* female flowers are often correlated with cylindrical spikes. This also holds for the other two species in the section. 5. *O. majorana* is related to *O. syriacum*, but differs from this species in its tomentellous indumentum, its usually obtuse leaves with veins not raised. 6. Some forms of *O. majorana* from Cyprus tend slightly towards *O. syriacum*. 7. Natural hybrids with *O. majorana* as a parent have not been found until now. Under cultivation or subnatural conditions *O. majorana* has formed hybrids with *O. vulgare* ssp. *virens* and ssp. *vulgare* (see pp. 138 and 134).

GREECE. NAXOS: 1844, *Lenormand s.n.* (G).

CYPRUS: between Panteleimon and Paleo Milo, on loam, c. 250 m, 24 May 1862, *Kotschy 937* (FI, JE). 1868, *Péronin s.n.* (G). Yalia, 24 May 1905, *Holmboe 798* (W). Aug. 1907, *Dunstan s.n.* (E). Lisso, c. 700 m, 10 – 16 June 1913, *Haradjian 886* (G, L). Ticco, c. 1500 m, 1 – 3 July 1913, *Haradjian 973* (G, L). Aphanis, c. 1200 m, *Kennedy 742* (G). Agios Hilarion, in dry rocky places, 7 June 1939, *Lindberg s.n.* (W). Yialousa, in dry rocky places, 21 June 1968, *Ecoyomides 1190* (G, AVU). Kambos, on igneous rocks, *Ecoyomides 1202* (AVU).

TURKEY. PROV. IÇEL: n. of Mersin, gorge of Guzel-Dere, 9 June 1855, *Balansa s.n.* (G). Mersin, 15 June 1896, *Peyron s.n.* (G). Oluçuk between Ermenek and Anamur, s. slopes, c. 1500 m, 18 Aug. 1949, *Davis 16333A* (G). C. 40 km w. of Anamur, in thick wood, c. 400 m, 4 June 1966, *Sorger 66 – 13 – 7* (Herb. Sorger). C. 60 km w. of Mut, in woods and on rocks, c. 700 m, 7 June 1966, *Sorger 66 – 30 – 8* (Herb. Sorger). C. 46 km s.e. of Mut, on rocky slope, c. 240 m, 19 June 1971, *Sorger 71 – 14 – 5* (Herb. Sorger).

FRANCE. CORSICA: Bastia, subspontaneous on old walls, 12 July 1868, *Debeaux s.n.* (FI, E).

ALGERIA: Oran, 8 June 1912, *Faure s.n.* (E).

MOROCCO: near Île de Tres Forcas, 12 June 1931, *Sennen & Mauricio 7985* (G).

SPAIN. ANDALUCIA: Malaga, in dry rocky places on limestone, c. 100 m, Aug. 1889, *Reverchon 538* (E, G, JE, W).

26. **Origanum onites** Linnaeus – **Figs. 19** and **20.**

O. onites Linnaeus, Sp. Pl.: 590 (1753); Sibthorp & Smith, Fl. Graeca 6: 58 (1826); Willkomm & Lange, Prodr. Fl. Hisp. 2: 399 (1868); Boissier, Fl. Or. 4: 553 (1879); Bonnet & Barratte, Cat. Pl. Vasc. Tunesie: 329 (1896); Halácsy, Consp. Fl. Graec. 2: 556 (1902); Holmes, Perf. Ess. Oil Rec. 4: 71 (1913); Post & Dinsmore, Fl. Syr. Palest. Sin. 2: 335 (1933); Thiebaut, Fl. Libano-Syr. 3: 47 (1953); Fiori, Nuova Fl. Anal. It. 2: 456 (1969); Tutin et al., Fl. Eur. 3: 172 (1972). *Majorana onites* (Linnaeus) Bentham, Lab. Gen. Sp.: 339 (1843); Gams, in Hegi, Ill. Fl. Mittel-Eur. 5: 2333 (1927); Hayek, Prodr. Fl. Penins. Balc. 2: 335 (1931); Rechinger, Fl. Aegaea: 532 (1943); Wolf, Baileya 2: 65 (1954); Briquet, Prodr. Fl. Corse 3: 220 (1955). – Type: *Linnaeus 743.11* (holo. LINN).

O. smyrnaeum Linnaeus, Sp. Pl.: 589 (1753). *Majorana smyrnaea* (Linnaeus) Kosteletzky, Med.-Pharm. Fl. 3: 769 (1834). *Schizocalyx smyrnaeus* (Linnaeus) Scheele, Flora, Neue Reihe 1: 575 (1834). – Type: *Linnaeus s.n.* (holo. BM).

Majorana cretica Miller, Gard. Dict. Abr. IV Ed.: 829 (1754).

O. album Salisbury, Prodr. Stirp.: 85 (1796).

O. pallidum Desfontaines. Cat. Pl. Horti Regii Paris.: 395 (1829).

Onites tomentosa Rafinesque, Fl. Tell. 3: 86 (1836).

O. tragoriganum Zuccagni ex Steudel, Nomencl. Bot.: 227 (1841).

O. orega Vogel, *Linnaea 15:* 78 (1841). *Majorana orega* (Vogel) Briquet, in Engler & Prantl. Nat. Pflanzenfam. 4(3a): 307 (1895).

Majorana oreja Walpers, Rep. Bot. Syst. 3: 697 (1844).

Majorana onites (Linnaeus) Bentham var. *columnaris* Rechinger, Denkschr. Akad. Wiss. Wien, Math.-Nat. Kl. 105(2): 127 (1943). – Type: *Rechinger 13948* Greece, Dia, (holo. W, iso. BM, G).

Subshrubs. Roots up to 1 cm in diameter. Young shoots hirsute. Stems erect or ascending, sometimes ramified at the bases, up to 100 cm long, light brown, hirsute (hairs c. 1.5 mm long), and glandular pilose. Branches of the first order present in the upper $\frac{1}{10}$ – $\frac{1}{5}$ of the stems, up to 8 pairs per stem, 2.5 (0.5 – 7.5) cm long; branches of the second order often present, those of the third order sometimes so. *Leaves* up to 28 pairs per stem, the lower ones shortly petiolate (petioles up to 6 mm long), heart-shaped, ovate or oval, tops ± acute or acuminate, margins often remotely serr(ul)ate, 14 (3–22) mm long, 12 (2–19) mm wide, hirsute (hairs c. 1 mm long) and glandular pilose, sessile glands up to 1700 cm^2; veins somewhat raised at the undersides. *Spikes* arranged in a false corymb, (sub)globose, ovoid or quadrigonus-cylindrical, 5 (3–17) mm long, c. 4 mm wide. *Bracts* 8 (4–34) pairs per spike, ovate, oval or obovate, tops acute, acuminate or obtuse, entire (sometimes denticulate), 3 (2–5) mm long, 2 (1.5–4) mm wide, light green, outside hairy. *Calyces* 1-lipped for c. $\frac{9}{10}$, ovate, obovate or somewhat rhomboid, 2.5 (2–3) mm long, outside pilosellous; upper lips entire or denticulate. *Corollas* 2-lipped for c. $\frac{2}{5}$, 4.5 (3–7) mm long (in female flowers c. 4 mm long), white, outside somewhat pilosellous; upper lips divided, for c: $\frac{1}{10}$–$\frac{1}{5}$, into 2, c. 0.2 mm long lobes; lower lips divided, for c. $\frac{4}{5}$, into 3 subequal, 1.5 (1.0–2.0) mm long lobes. *Stamens* protruding; filaments up to 4 and 5 mm long. *Styles* up to 10 mm long. *Chromosome number* 2n = 30.

Geography and ecology. *O. onites* has a rather large distribution area: it is found in southern Greece, on Kriti and many other Greek islands, and in western and southern Turkey. One isolated site is known from Sicily, near Siracusa. Other records probably concern garden escapes. *O. onites* usually grows on slightly elevated rocky places, from sea level up to 1400 m, often on limestone. It flowers from April to August.

Notes. 1. *O. onites* differs from *O. syriacum* in its hirsute, often serrate leaves and its corymbiform inflorescences. 2. Hybrids have been found from *O. onites* and *O. sipyleum* and from *O. onites* and *O. vulgare* ssp. *hirtum* (see pp. 137 and 136).

TURKEY. PROV. IÇEL: near ruins c. 27 km n. of Silifke, c. 10 m, 18 June 1971, *Sorger 71–13–17* (Herb. Sorger). C. 36 km n. of Silifke, c. 100 m, 22 June 1971, *Sorger 71–27–5* (Herb. Sorger). PROV. ANTALYA: near Elmali, on stony hills, 17 June 1860, *Bourgeau 219* (G, JE). Near Perge, c. 100 m, 22 May 1963, *Sorger 63–25–20* (Herb. Sorger). C. 13 km w. of Antalya, c. 10 m, 26 June 1965, *Sorger 65–33–65* (Herb. Sorger). Elmalidağ, s. exposed slope on steppe, 1200–1400 m, 23 June 1967, *Sorger 67–21–5* (Herb. Sorger). PROV. ISPARTA: w. shore of Beyşehir lake, open *Juniperus* wood, c. 1150 m, 15 June 1966, *Sorger 66–48–9* (Herb. Sorger). PROV. DENIZLI: near Denizli, 7 July 1905, *Saint-Lager s.n.* (G, L). Between Motel Mistur and the road to Denizli, on rocky slope, 450–550 m, 28 June 1973, *Buttler & Erben 17513* (Herb. Buttler). PROV. IZMIR: Izmir, along roads, 24 June 1854, *Balansa 317* (G, JE, W). Near Izmir, on hills, 29 May 1906, *Bornmüller 9855* (BM, G, JE, W). Nif Dağ s. of Kemâlpaşa, in pine wood, c. 980 m, 23 June 1973, *Buttler & Erben 17345* (Herb. Buttler). PROV. MANISA: near Manisa, 25 June 1905, *Saint-Lager s.n.* (G).

GREECE. RHODOS: near Bastida, on uncultivated hills, 28 May 1870, *Bourgeau 139* (E, W). Near Filermos, 30 May 1938, *Engelhardt s.n.* (JE). SAMOS: near Vathi, on schist, 16–23 June 1932, *Rechinger 1919* (BM). LESBOS: Mytilene, in open rocky places, 20–21 May 1927, *Rechinger 1212* (W). Apr. 1968, *Bedford s.n.* (BM). KARPATHOS: 18 May 1866, *Forsyth Major 203* (G). Near Vrondi opposite Pigadia, rocky places (limestone) beneath pine trees, 17 June 1935, *Rechinger 8255* (BM). Near Holethria e. of Calilimi, 22 July 1950, *Davis 18085* (W). KRITI: in rocky places near cave Melidonis, 4 Aug. 1893, *Baldacci 165* (BM, W). Sitia, between Sphaka and Turloti, in rocky places (limestone), c. 300 m, 17 May 1942, *Rechinger 13024* (BM, G). DIA: in rocky places, 30 May 1899, *Baldacci 41* (BM, G, W). Near bay Panagia, in rocky places (limestone), 23 June 1942, *Rechinger 13948* (W, BM, G). THIRA (SANTORINI): in rocky places, July 1880, *Letourneux 355* (W). Ios: in rocky places, common, 29 June 1889, *Heldreich s.n.* (BM, E, JE, W). NAXOS: July 1897, *Leonis s.n.* (G, W). POROS: 4 June 1928, *Guiol 268* (BM). JURA (GYAROS): June 1896, *Leonis s.n.* (JE). On calcareous rocks, 10 – 12 May 1927, *Rechinger 1054* (W). KYRIA PANAGIA: on calcareous rocks, 10 – 12 May 1927, *Rechinger 1014* (W). MAINLAND: near Navplion, 28 Apr. 1849, *Orphanides 143* (BM, JE, W). Peninsula Methanon, on rocks, 1885, *Haussknecht s.n.* (BM, JE). Peninsula Methanon, 23 May 1885, *Heldreich s.n.* (W). Near Navplion, 27 July 1899, *Saint-Lager s.n.* (G). Between Phronia and Aria near Navplion, 25 June 1901, *Saint-Lager s.n.* (G). Near Argos, 12 July 1906, *Tunta s.n.* (JE). Near Mycenae, 20 May 1929, *Guiol s.n.* (BM). Peninsula Malea, between Daimonia and Monemvasia, in rocky places (limestone), 7 – 9 June 1958, *Rechinger 20077* (G).

TUNISIA: valley of l'Oued Bou-Dissar near Aïn Zraris, 12 June 1888, *Cosson s.n.* (G).

ITALY. SICILY: near Siracusa, in dry calcareous hills, June 1894, *Ross 64* (BM, G, L). Siracusa, in the former amphitheatre, 16 May 1927, *Ronniger s.n.* (W).

27. **Origanum syriacum** Linnaeus

O. syriacum Linnaeus, Sp. Pl.: 590 (1753). *Amaracus syriacus* (Linnaeus) Stokes, Bot. Mat. Med. 3: 347 (1812). *Majorana syriaca* (Linnaeus) Kosteletzky, Med.-Pharm. Fl. 3: 768 (1834). *Schizocalyx syriacus* (Linnaeus) Scheele, Flora, Neue Reihe 1: 575 (1843). – Type: *Linnaeus 743.12* (holo. LINN).
O. aegyptiacum auct. non Linnaeus; Savi, Osserv. Gen. Origanum: 3 (1840). *Majorana aegyptiaca* (auct. non Linnaeus) Kosteletzky, Med.-Pharm. Fl. 3: 770 (1834).
Majorana scutellifolia Stokes, Bot. Mat. Med. 3: 349 (1812).
Other synonyms are cited for each of the three varieties, which are recognized.

Subshrubs. Roots up to 1 cm in diameter. Young shoots (densely) tomentose or hirsuto-tomentose. Stems ascending or erect, often ramified at the bases, up to 90 cm long, (light) brown, tomentose or hirsute (hairs c. 1.5 mm long) and somewhat glandular pilose. Branches of the first order present, in the upper $\frac{1}{10}$ – $\frac{4}{5}$ of the stems, up to 15 pairs per stem, 2.5 (0.4 – 13) cm long; branches of the second order usually present, those of the third order often so. *Leaves* up to 30 pairs per stem, clearly petiolate to (sub)sessile (petioles up to 8 mm long), ovate, oval or heart-shaped, tops obtuse to acuminate, margins entire or remotely crenulate or serr(ul)ate, 17 (3 – 35) mm long, 11 (2 – 23) mm wide, green or whitish, slightly hirsuto-tomentose to densely tomentose (hairs c. 1 mm long), sessile glands up to 900 per cm^2; veins usually raised at the underside. *Spikes* quadrigonus-cylindrical or subglobose, 7 (3 – 25) mm long, c. 4 mm wide. *Bracts* 9 (4 – 40) pairs per spike, obovate or oval, tops obtuse or acute, entire or slightly denticulate, 2.5 (2 – 5) mm long, c. 2 mm wide, green or whitish, outside (hirsuto-)tomentose. *Calyces* 1-lipped for $\frac{9}{10}$ or more, obovate or oval, c. 2 mm long, outside (hirsuto-)tomentose; upper lips entire, or slightly denticulate or lobate. *Corollas* 2-lipped for c. $\frac{2}{5}$, 5 (4 – 7.5) mm long, white, outside more or less pilosellous; upper lips divided, for c. $\frac{1}{5}$, into 2, c. 0.2 mm long lobes; lower lips divided for c. $\frac{4}{5}$, into 3 subequal, 1.6 (1.0 – 2.3) mm long lobes. *Stamens* protruding; filaments up to 4 and 5 mm long. *Styles* up to 9 mm long.

Geography and ecology. *O. syriacum* inhabits a large area in the eastern Medittetranean. It is found in southern Turkey, on Cyprus, in Syria, Lebanon, Israel, Jordan and on the Sinai Peninsula, and grows from nearly sea-level up to at least 2000 m. It grows in rocky soils, often on limestone, and it flowers from May to October.

Notes. 1. The three species in the section *Majorana* are rather closely related. All possess the same type of spikes, bracts, calyces and corollas. The main differences lie in the shape of the inflorescences and in the indumentum. So *O. syriacum* differs from *O. majorana* in its hirsute or tomentose stems and its more or less tomentose leaves with usually acute tops and raised veins on the under side. From *O. onites* it differs in its paniculate (not corymbiform) inflorescences. 2. No less than four natural hybrids are known with *O. syriacum* as a parent. The other parental species are: *O. libanoticum*, *O. bargyli*, *O. ehrenbergii* and *O. laevigatum* (see pp. 133, 139 135 and 140). 3. Three varieties are recognized mainly based on differences in the indumentum and the leaves.

a. var. **syriacum** – **Figs. 20** and **21.**

O. syriacum Linnaeus; Holmboe, Stud. Veg. Cyprus: 162 (1914); Thiebaut, Fl. Libano-Syr. 3: 47 (1953).
O. maru Linnaeus, Sp. Pl. II Ed. 2: 825 (1763); Holmes, Perf. Ess. Oil Rec. 4: 69 (1913); Post & Dinsmore, Fl. Syr. Palest. Sin. 2: 334 (1933). *Majorana maru* (Linnaeus) Briquet, in Engler & Prantl, Nat. Pflanzenfam. 4(3a): 307 (1895). – Type: *Linnaeus 743.12* (holo. LINN).
Majorana crassa Moench, Meth. Pl.: 406 (1794). *O. crassa* (Moench) Chevallier, Rev. Bot. Appl. 18: 597 (1938).
O. vestitum Clarke, Trav. 2: 451 (1823).
Majorana crassifolia Bentham, Lab. Gen. Sp.: 339 (1834); Walpers, Rep. Bot. Syst. 3: 696 (1844); Chevallier, Rev. Bot. Appl. 18: 596 (1938). – Type: *Sieber s.n.*, Israel, Jerusalem (holo. PRC).
O. maru Linnaeus var. *capitatum* Post, Fl. Syr. Palest. Sin.: 617 (1896).

Stems tomentose. *Leaves* shortly petiolate or subsessile, usually ovate or oval, 12 (3 – 22) mm long, 7 (2 – 14) mm wide, whitish, tops obtuse or acute, margins entire. *Spikes* quadrigonus-cylindrical or subglobose, 8 (4 – 20) mm long. *Bracts* 2.5 (2 – 3.5) mm long, c. 2 mm wide, whitish. *Calyces* c. 2 mm long. *Corollas* 5 (4 – 6) mm long.

Israel: mountain range in the environs of Nazareth, 1832, *Bové s.n.* (G). Environs of Jerusalem, Aug. 1881, *Burdet 328* (G). Mountains of Judaea, near Jerusalem, 30 July 1887, *Burdet 15* (G). Mt. Ebal, 10 Aug. 1887, *Burdet 60* (G). Bethlehem, May 1889, *Jouannet-Marie s.n.* (W). Judaea, in calcareous mountains, near Bab-el-Wad, 16 May 1897, *Bornmüller 1247* (JE). Samaria, rocky places, c. 450 m, 11 Aug. 1912, *Meyers & Dinsmore B4037* (BM, G). Kiriath Anavim near Jerusalem, 26 May 1931, *Zohary & Amdursky 165* (BM, E, G, L, U).
Jordan: Jerash, 1928, *Crowfoot 84* (BM). Jerash, 21 Sept. 1946, *Lupton s.n.* (BM).
Syria: Lattakia, June 1907, *Vetters s.n.* (W).

b. var. **bevanii** (Holmes) Ietswaart

O. syriacum Linnaeus var. *bevanii* (Holmes) Ietswaart, *stat. nov. O. bevanii* Holmes, Perf. Ess. Oil Rec. 6: 19 (1915), *("bevani")*. *Majorana bevani* (Holmes) Chevallier, Rev. Bot. Appl. 18: 596 (1938). – Type: *Bevan s.n.*, Cyprus, near Lapithos (holo. K).
O. maru L. f. *viridula* Bornmüller, Herb. exsicc. Iter Syr.: no. 1248 (1897), *nomen nudum*. – Type: *Bornmüller 1248*, Lebanon, Lebanon Mts, near Brummana (holo. G, iso. JE, L, W).
O. pseudo-onites Lindberg, Acta Soc. Sc. Fenn., Nova Ser. B, 2: 29 (1946). – Type: *Lindberg s.n.*, Cyprus, near Lapithos (holo. S, iso. W, K).

Stems hirsute. *Leaves* petiolate, usually more or less ovate, rather large, 25 (5 – 35) mm long, 15 (4 – 23) mm wide, greenish, slightly tomentose, tops obtuse or more or less acute, margins entire or remotely serrulate or crenulate. *Spikes* subglobose or quadrigonus-cylindrical, 8 (5 – 25) mm long. *Bracts* 3 (2 – 5) mm long, 2 (1.5 – 3.5) mm wide, greenish. *Calyces* c. 2.5 mm long. *Corollas* 6 (4.5 – 7.5) mm long.

TURKEY. PROV. IÇEL: Taurus Mts., summer 1836, *Kotschy s.n.* (W). PROV. ADANA: near Karatepe, c. 200 m, 23 June 1971, *Sorger 71 – 31 – 28* (Herb. Sorger). PROV. HATAY: Amanus Mts., near Beilan, near aquaduct in valley Dary Derre, c. 1000 m, 21 June 1862, *Kotschy 48* (JE, L, W). Amanus Mts., c. 1000 m, Aug. 1906, *Haradjian 451* (E, G, W). Amanus Mts., near Egby, c. 500 m, 1906, *Haradjian 668* (E, G). Amanus Mts., Küsliji Dagh, Aug. 1908, *Haradjian 2449* (E, G, W). Amanus Mts., c. 800 m, Sept. 1913, *Haradjian 4644* (E, G, W). W. of Belen, on limestone rock and debris, c. 500 m, 15 June 1953, *Huber-Morath 12708* (Herb. Huber-Morath).

CYPRUS: near Lapithos, 10 June 1913, *Bevan s.n.* (type). Kyrenia, near Lapithos, on dry slopes, 13 June 1939, *Lindberg s.n.* (W).

SYRIA: 1839, *Aucher s.n.* (W).

LEBANON: spurs of Lebanon Mts., n.e. of Saïda, 24 May 1853, *Blanche 86* (JE, W). Lebanon Mts., near Bscherre, 19 July 1855, *Kotschy 260* (BM, W). Environs of Brummana, on slope w. of the chain of the Lebanon Mts., July-Oct. 1879, *Cramer s.n.* (G). Near Brummana, lower regions of the Lebanon Mts., in pine woods, c. 750 m, June 1879, *Bornmüller 1248* (G, JE, L, W). Lebanon Mts., near monastry of Becharré, in bush of *Quercus ilex*, c. 1350 m, 2 July 1931, *Zerny s.n.* (W). Below Sû, 17 June 1943, *Davis 6439A* (E).

c. var. **sinaicum** (Boissier) Ietswaart

O. syriacum Linnaeus var. *sinaicum* (Boissier) Ietswaart *comb. nov.* *O. maru* Linnaeus var. *sinaicum* Boissier, Fl. Or. 4: 553 (1879). – Type: *Bové s.n.*, Sinai Peninsula, Mt. Sinai (holo. G, iso. L, W).

Majorana nervosa Bentham, Lab. Gen. Sp.: 339 (1834). *O. nervosum* (Bentham) Vogel, Linnaea 15: 78 (1841); Chrtek, Acta Univ. Car., Ser. Biol. 1968: 239 (1969). *Majorana maru* (Linnaeus) Briquet var. *nervosa* (Bentham) Briquet, in Engler & Prantl, Nat. Pflanzenfam. 4(3a): 307 (1895). – Type: *Bové s.n.*, Sinai Peninsula, Mt. Sinai (holo. G, iso. L, W).

O. maru Linnaeus var. *aegyptiacum* (auct. non Linnaeus) Dinsmore, Pl. Post. Dinsm. 1: 11 (1932). *O. syriacum* Linnaeus var. *aegyptiacum* (auct. non Linnaeus) Täckholm, Stud. Fl. Egypt: 143 (1956).

Stems hirsute. *Leaves* more or less petiolate, heart-shaped or ovate, 14 (4 – 25) mm long, 11 (3 – 20) mm wide, greenish, slightly hirsuto-tomentose, tops obtuse to acuminate, margins entire or remotely crenulate or serrulate. *Spikes* usually subglobose sometimes quadrigonus-cylindrical, 5 (3 – 15) mm long. *Bracts* 2.5 (2 – 3) mm long, c. 1.5 mm wide, greenish. *Calyces* c. 2 mm long. *Corollas* 4 (4 – 6) mm long.

SINAI PENINSULA: environs of Mt. Sinai, June 1832, *Bové s.n.* (type). Mt. Sinai, on rocks, 13 June 1835, *Schimper s.n.* (E, G, L, W). Ouadi Tarfa, s. of monastery, 7 May 1891, *Cramer s.n.* (G, W). Gebel Serbâl 1892, *Grote s.n.* (JE).

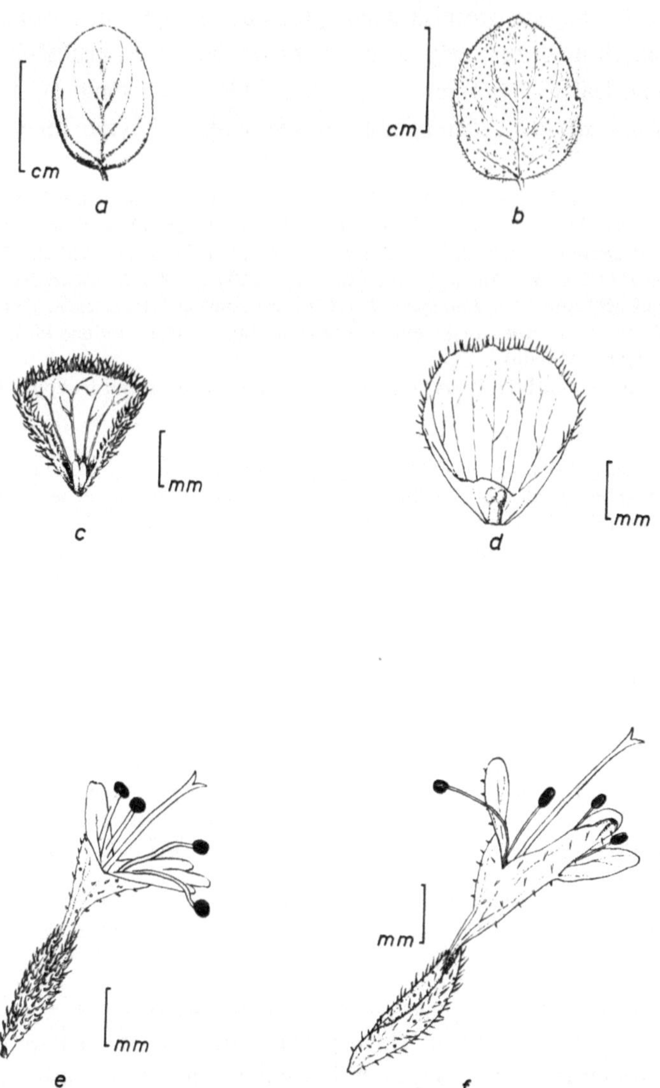

Figure 19. Leaves, calyces in lower lip view, and flowers with bracts in side view of the species in the section *Majorana*, except *O. syriacum* (for which see figure 21): a., c. and e. *O. majorana;* b., d. and f. *O. onites*.

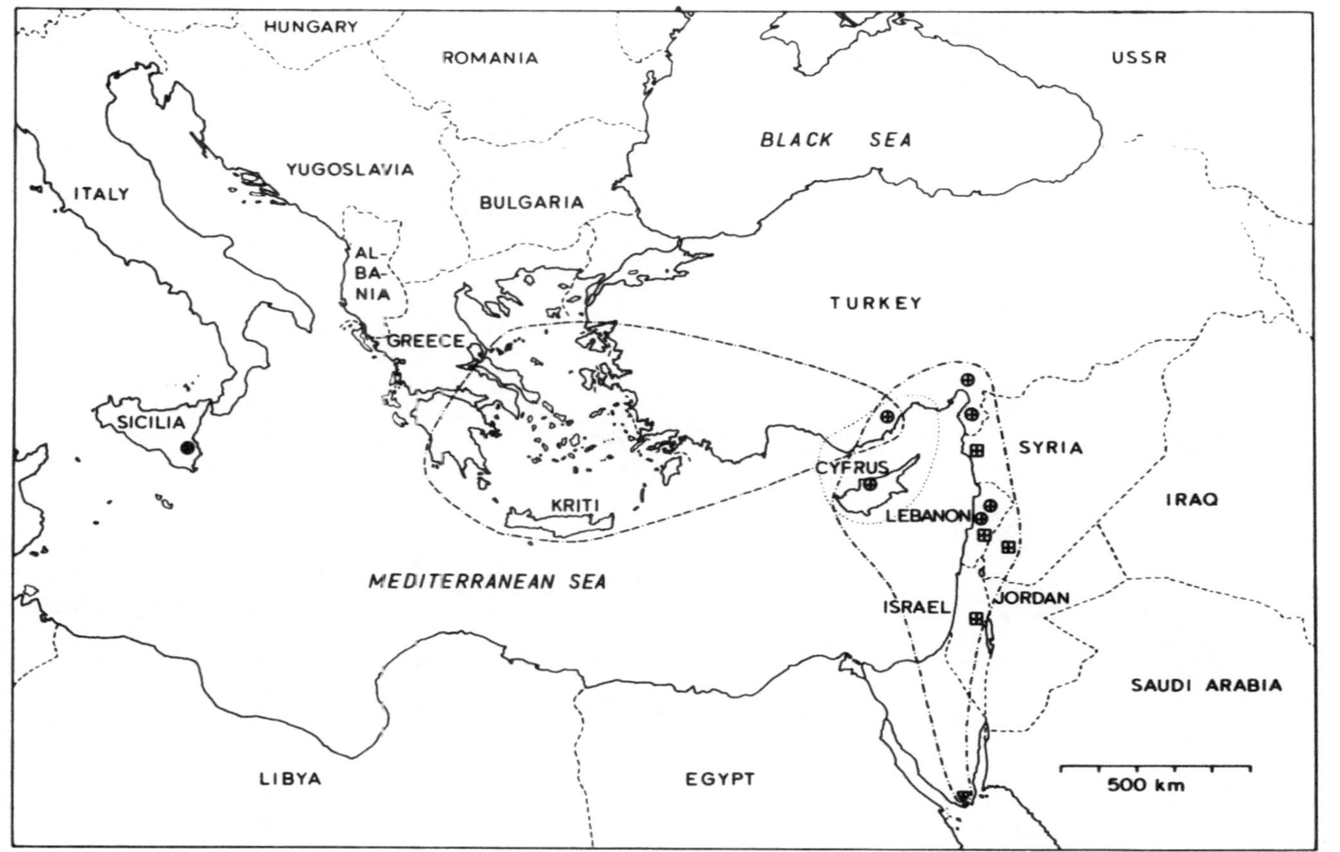

Figure 20. Distribution of the species and varieties in the section *Majorana*: .... *O. majorana*; —.—.—.— and ● *O. onites*; —.—.— *syriacum*, and its varieties: ⊕ var. *bevanii*; ⩔ var. *sinaicum*; ⊞ var. *syriacum*.

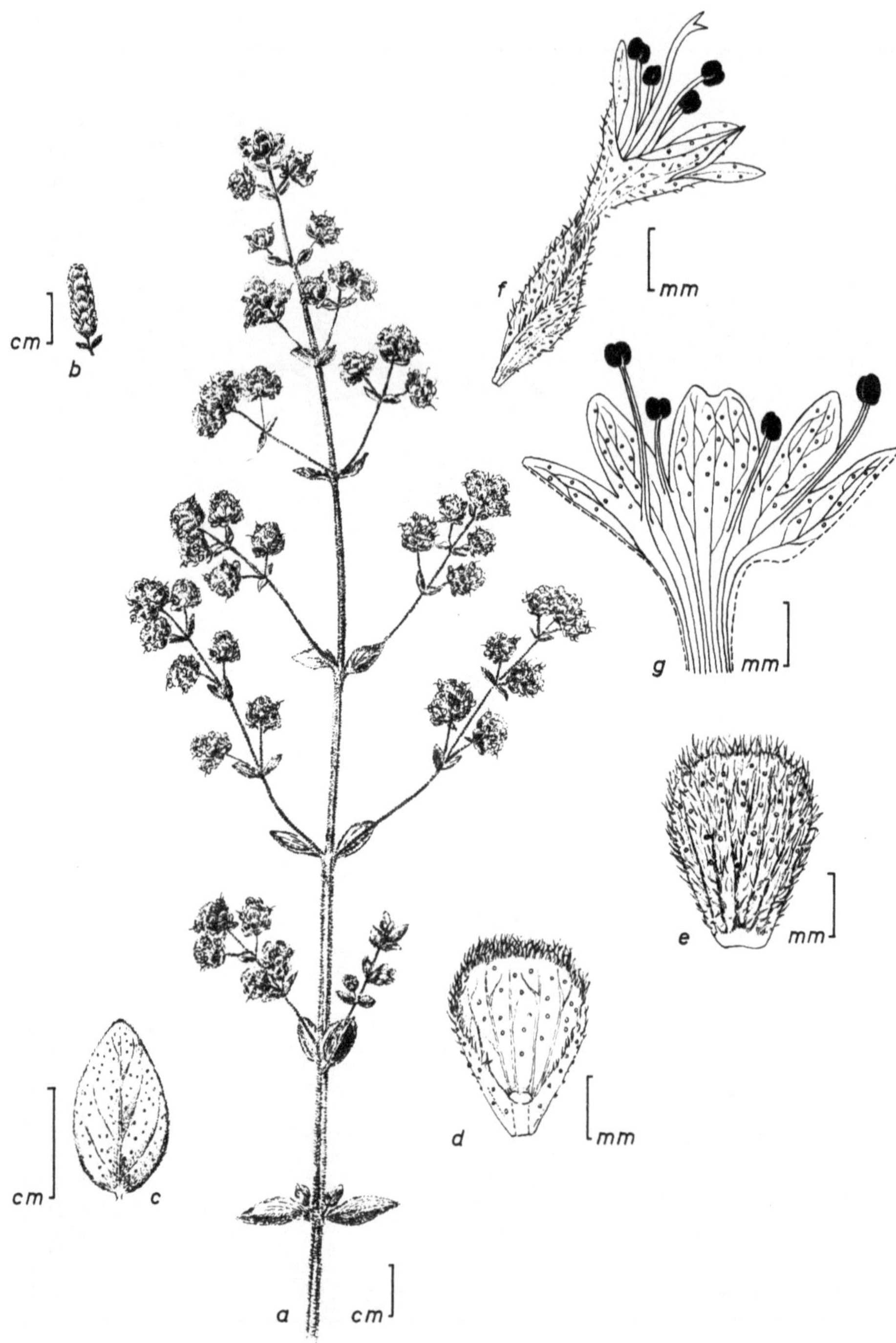

Figure 21. O. syriacum var. *syriacum:* a. habit; b. quadrigonus-cylindrical spike; c. leaf; d. calyx in lower lip view; e. bract outside; f. flower with bract in side view; g. corolla cut through the lower lip.

VII. **Section Campanulaticalyx** Ietswaart

Section *Campanulaticalyx* Ietswaart *sect. nov.* – Type: *Origanum dayi* Post.

Rami primarii vulgo, secundarii raro adsunt. Folia herbacea vel plus minusve coriacea. Spicae satis laxae, caulibus et ramis foliatis aliquanto similes, parvae, erectae. Bracteae foliis aliquanto parviores, eis similes textura coloreque, paulo vel paene imbricatae, calycibus $2\frac{1}{2}$plo breviores vel longitudine aequantes. Flores duo pro verticillastro, bisexuales vel feminei. Calyces (tubuloso-) campanulati (etiam statu frugifero), dentibus subaequalibus vel aequalibus, usque ad circa duas quintas partes divisi; fauces pilosae. Corollae bilabiatae, usque ad circa duas quintas partes vel usque ad circa quartam partem incisae, calycibus circa $1\frac{1}{2}$ – 2plo longiores. Stamina aliquanto inaequalia, stricta, longe exserta vel subinclusa; filamentis corollas plus minusve longitudine aequantibus, pilosellis vel glabris.

Branches of the first order usually present, those of the second order seldom so. *Leaves* herbaceous or somewhat leathery. *Spikes* rather loose, not clearly distinct from the leaved stems and branches, small, erect. *Bracts* somewhat smaller than the leaves, but leaflike in texture and colour, slightly imbricate or nearly not so, $\frac{2}{5}$ – 1 x calyces. *Flowers* 2 per verticillaster, usually bisexual, very small or medium sized. *Calyces* (tubular-)campanulate (even when fruit bearing), with 5 (sub)equal teeth for c. $\frac{2}{5}$; throats pilose. *Corollas* 2-lipped for $\frac{2}{5}$ or $\frac{1}{4}$, c $1\frac{1}{2}$ – 2 × calyces. *Stamens* slightly unequal in length, straight, far protruding or (sub)included; filaments ± as long as corollas, pilosellous or glabrous.

28. **Origanum dayi** Post – **Figs. 22** and **23.**

O. dayi Post, Bull. Herb. Boiss. 1: 405 (1893); Zohary, Fedde Rep. 28: 63 (1930); Post & Dinsmore, Fl. Syr. Palest. Sin. 2: 333 (1933); Danin, Israel J. Bot. 16: 101 (1967). – Type: *Post 244*, Israel, between Hebron and Zuweirat-el Foqa (holo. G, iso. BM).

Satureja camphorata Bornmüller, Mitt. Thür. Bot. Ver., Neue Folge 30: 80 (1913). – Type: *Bornmüller B1023*, Israel, Judea, e. of Carmel (holo. B).

Subshrubs, flowers bisexual and possibly also female only. Stems up to 70 cm long, erect, basely ramified or not, light yellow brown, ± hirsute (hairs c. 1.5 mm long) and glandular pilose. Branches of the first order always present, in the upper $\frac{1}{2}$ of the stems, up to 30 pairs per stem, 1.5 (0.5 – 4) cm long, usually not ramified. *Leaves* up to 40 pairs per stem, sessile, heart-shaped or ovate, tops acute or ± acuminate, thin, light green, glandular pilose and somewhat pilose (hairs c. 0.7 mm long, mainly at the margins), sessile glands up to 2500 per cm^2; those of stems and non-flowering branches 8 (6 – 12) mm long and 6 (4 – 8) mm wide, those of flowering branches shading off into the bracts, 5 (2 – 6) mm long and 3 (2.5 – 4) mm wide. *Spikes* cylindrical, 12 (8 – 30) mm long, c. 6 mm wide. *Bracts* 6 (2 – 20) pairs per spike, ovate or oval, tops ± acute, c. 4 mm long, c. 2.5 mm wide, somewhat pilose. *Flowers* with c. 1 mm long pedicels. *Calyces* campanulate, with 5 subequal teeth, toothed for c. $\frac{2}{5}$, 6 (4.5 – 7.5) mm long, outside pilose; teeth 2.5 (1 – 3) mm long. *Corollas* 2-lipped for c. $\frac{1}{4}$, 9 (7 – 11) mm long, whitish, outside pilosellous; upper lips divided, for c. $\frac{1}{10}$, into 2, c. 0.2 mm long lobes; lower lips divided, for c. $\frac{2}{5}$, into 3 subequal, c. 1 mm long lobes. *Stamens* far protruding; filaments pilosellous, up to 10 and 10.5 mm long. *Styles* up to 16 mm long.

Geography and ecology. *O. dayi* occurs in the Judean desert and the northern Negev, up to 800 m. It is found in crevices of hard limestone and dolomites, and on flint flats. It also grows in "wadis" adjacent to these rocks. In the Judean desert it grows in batha vegetations together with amongst others *Thymelaea hirsuta* and *Iris palestina.* It has been found flowering in July and August.
Notes. 1. In the specimens studied far protruding stamens were found, while Zohary (1930) stated that the stamens are included. Possibly the species must be considered as gynodioec. 2. *O. dayi* and *O. ramonense* are the only two species in the genus *Origanum* possessing pilosellous staminal filaments. 3. *O. dayi* differs from *O. ramonense*, to which it is related, in its longer and villous stems, its longer campanulate calyces and its whitish corollas. From *O. isthmicum* it differs in its villous stems, and larger flowers with far protruding stamens.

ISRAEL: between Hebron and Zuweirat-el Foqa, on sunny hills, 22 Aug. 1892, *Post 244* (type). S.e. of Carmel, 16 Febr. 1912, *Meyers B1023* (B, G). Judea, e. of Carmel, crevices of rocks, 16 Febr. 1912, *Bornmüller B1023* (B).

29. **Origanum isthmicum** Danin – **Figs. 23** and **24.**

O. isthmicum Danin, Israel J. Bot. 18: 191 (1969). – Type: *Danin s.n.*, North Sinai, Isthmic Desert, Wadi Abu Seiyal (holo. HUJ, iso. AVU).

Subshrubs, flowers bisexual. Stems up to 50 cm long, much branched, the older twigs greyish, the younger (light) brown, usually somewhat pilosellous (hairs c. 0.2 mm long, mostly on the more slender twigs; on nodes sometimes also a few c. 1 mm long hairs present). Branches of the first order present, 1.5 (0.5 – 5) cm long, sometimes branches of the second order also present. *Leaves* (sub)sessile, (broadly) heart-shaped or roundish, tops usually obtuse, somewhat leathery, glaucous, slightly pilosellous (hairs c. 0.1 mm long, especially on the leaves of the flowering branches), sessile glands up to 1000 per cm^2; those of stems and non-flowering branches 5 (3 – 7) mm long, 5 (3 – 7) mm wide, those of flowering branches shade off into the bracts, c. 2 mm long and wide. *Spikes* cylindrical or ovoid, 5 (2 – 20) mm long, c. 3 mm wide. *Bracts* 3 (2 – 8) pairs per spike, heart-shaped or roundish, tops obtuse or ± acute, c. 2 mm long and wide, pilosellous. *Flowers* subsessile. *Calyces* tubular-campanulate, with 5 (sub)equal teeth, toothed for c. $\frac{2}{5}$, 2 (1.5 – 3) mm long, outside pilosellous; teeth c. 1 mm long. *Corollas* 2-lipped for c. $\frac{2}{5}$, 3 (2.5 – 3.5) mm long, yellowish white, pilosellous; upper lips divided, for c. $\frac{2}{5}$, into 2, c. 0.5 mm long lobes; lower lips divided, for c. $\frac{4}{5}$, into 3 subequal, c. 1 mm long lobes. *Stamens* (sub)included; filaments up to c. 2 mm long. *Styles* up to 3.5 mm long.

Geography and ecology. *O. isthmicum* has been discovered recently in the Isthmic desert in the northern part of the Sinai Peninsula. Here it grows at c. 500 m in crevices of hard limestone, at northern exposure. The rainfall in this area is no more than c. 100 mm per year. According to Danin *O. isthmicum* should be considered as a relic endemic species of a former mesic flora. It has been found flowering in June. Notes. 1. Owing to its many branched stems some diagnostic characters mentioned in all other species (e.g. number of branches and leaves per stem) are omitted. 2. *O. isthmicum* holds a rather isolated position within the section *Campanulaticalyx*, on behalf of its very small flowers with subincluded stamens and styles.

NORTH SINAI: Isthmic desert, c. 55 km s.e. of El A'rish, Gebel Halal, Wadi Abu Seiyal, in crevices of hard limestone, c. 500 m, n. exposure, 3 Apr. 1968, *Danin & Tadmor s.n.* (AVU, HUJ). Ibid., 18 June 1968, *Danin s.n.* (type).

30. Origanum ramonense Danin – Figs. 23 and 24.

O. ramonense Danin, Israel J. Bot. 16: 101 (1967). – Type: *Danin 13141*, Israel, Central Negev, Nahal Eilot (holo. HUJ, iso. AVU).

Subshrubs, flowers bisexual. Stems up to 40 cm long, ramified more or less at the bases, greyish, piloso-tomentose (hairs c. 0.8 mm long, but length varying), and glandular pilose. Branches of the first order present, in the upper $\frac{3}{5}$ of the stems, up to 25 pairs per stem, 0.3 (0.1 – 0.7) cm long, usually not ramified. *Leaves* up to 30 pairs per stem, (sub)sessile, heart-shaped, tops acute, thin, (light) green, piloso-tomentose (hairs c. 0.5 mm long, but length varying) and glandular pilose, sessile glands inconspicuous, up to 2000 per cm^2; those of the stems and non-flowering branches 6 (3 – 10) mm long and 5 (3 – 8) mm wide, those of flowering branches shade off into the bracts, c. 3 mm long and c. 2 mm wide. *Spikes* cylindrical or subglobose, 6 (3 – 10) mm long, c. 4 mm wide. *Bracts* 5 (2 – 8) pairs per spike, ovate or oval, tops ± obtuse, c. 2.5 mm long, c. 1.5 mm wide, piloso-tomentose. *Flowers* with c. 0.5 mm long pedicels. *Calyces* tubular-campanulate, with 5 subequal teeth, toothed for c. $\frac{2}{5}$, 4 (3.5 – 4.5) mm long, outside piloso-tomentose; teeth 1.2 (1 – 1.7) mm long. *Corollas* 2-lipped for c. $\frac{2}{5}$, 8 (6 – 10) mm long, purplish pink, outside pilosellous; upper lips divided, for c. $\frac{1}{10}$, into 2, c. 0.2 mm long lobes; lower lips divided, for c. $\frac{2}{5}$, into 3 subequal, c. 1.5 mm long lobes. *Stamens* far protruding; filaments pilosellous, up to 10.5 and 11 mm long. *Styles* up to 15 mm long.

Geography and ecology. *O. ramonense* has recently been described from Ramon Mts. in the Negev, where it occurs at 800 – 1000 m, on hard limestone. It has been found flowering in September.
Note. *O. ramonense* is related to *O. dayi*, from which it differs in its shorter, more branched, villoso-tomentose stems, its smaller tubular-campanulate calyces and purplish pink corollas.

CENTRAL NEGEV: Ramon Mts., Nahal Eilot, in crevices of hard limestone, c. 950 m., 27 Sept. 1966, *Danin 13141* (type). Central Negev, Nahal Lotz, in crevices of hard limestone, 22 March 1965, *Danin & Gravish s.n.* (HUJ, AVU).

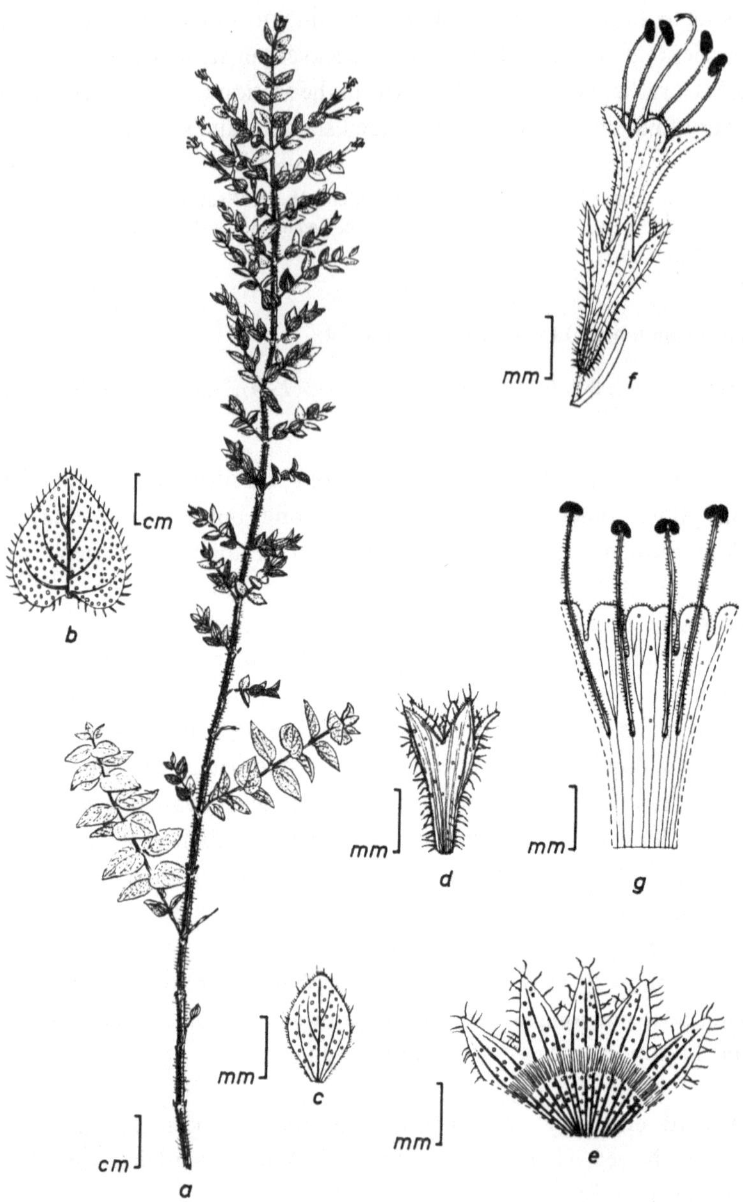

Figure 22. O. dayi: a. habit; b. leaf; c. bract; d. calyx; e. calyx cut through the lower lip; f. flower with bract in side view; g. corolla cut through the lower lip.

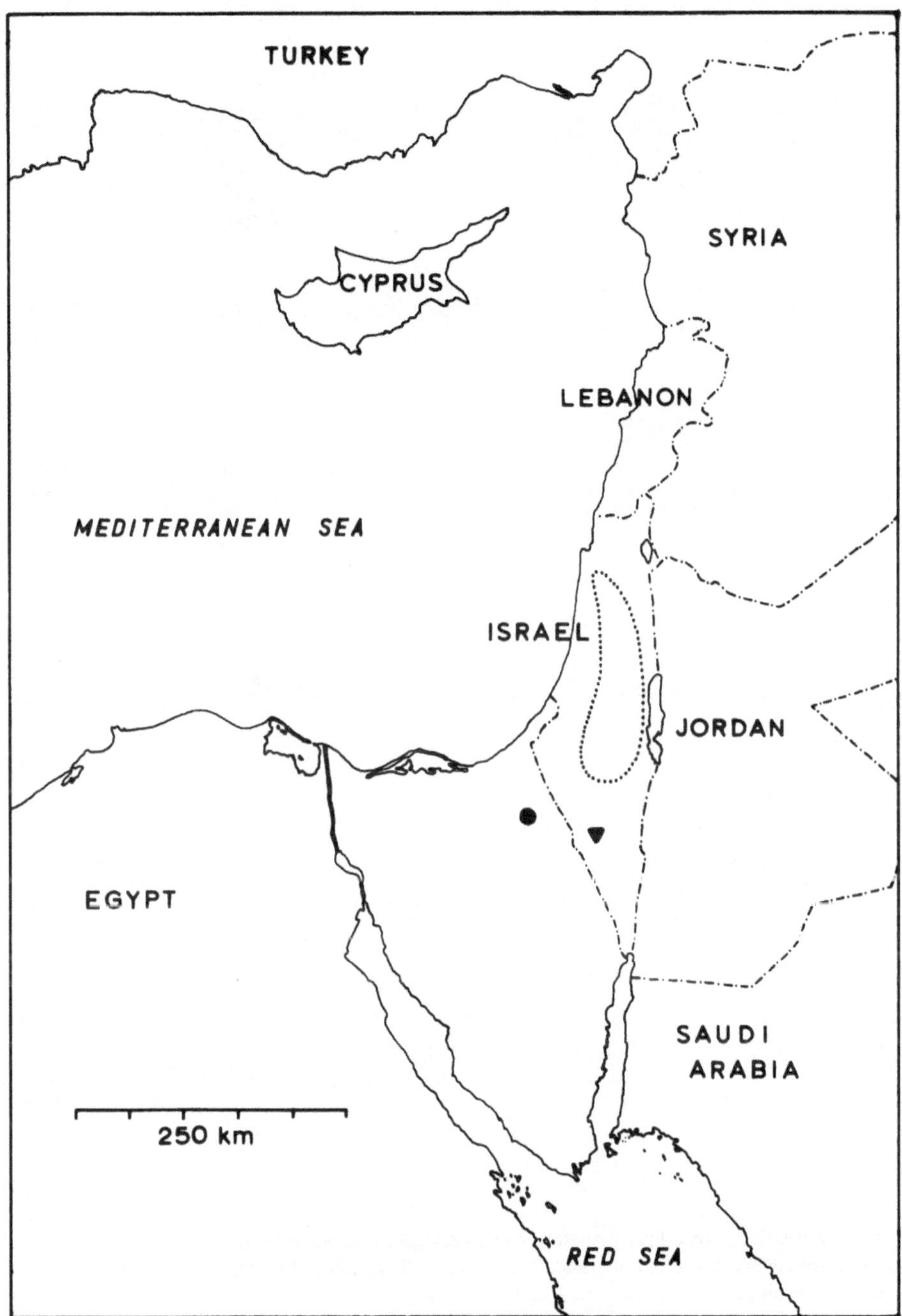

Figure 23. Distribution of the species in the section *Campanulaticalyx:* .... *O. dayi;* ● *O. isthmicum;* ▼ *O. ramonense.*

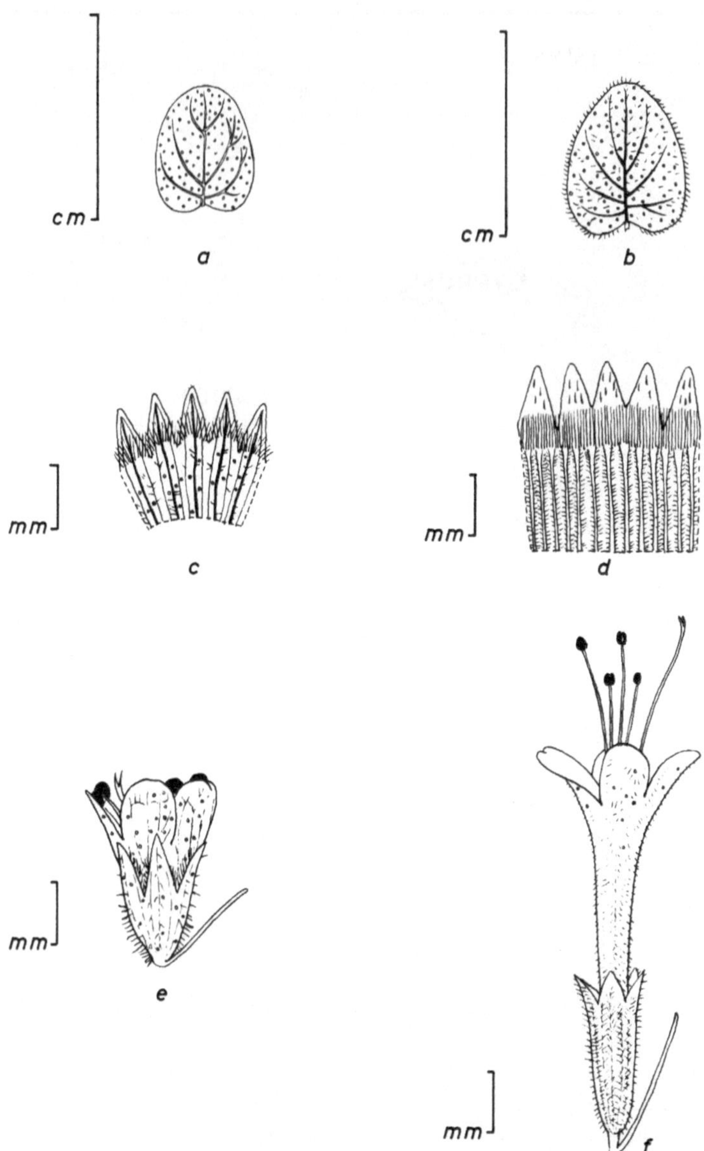

Figure 24. Leaves, calyces cut through the lower lip, and flowers with bracts in side view of the species in the section *Campanulaticalyx*, except *O. dayi* (for which see figure 22): a., c. and e. *O. isthmicum;* b., d. and f. *O. ramonense*.

VIII. Section Elongatispica Ietswaart

Section *Elongatispica* Ietswaart *sect. nov.* – Type: *Origanum floribundum* Munby.

Rami primarii semper, secundarii vulgo, tertiarii interdum adsunt. Folia herbacea vel paulo coriacea. Spicae laxae vel laxissimae, tenues, distinctae, longae vel longissimae, quaeque saepe omnino e unio solo ramo constantes. Bracteae foliis forma magnitudineque impares, textura coloreque eis similes, aliquanto vel haud imbricatae, calycibus $1\frac{1}{3}$plo breviores usque ad $1\frac{1}{3}$plo longiores. Flores duo pro verticillastro, bisexuales vel feminei, minuti vel minutissimi. Calyces plus minusve tubulares, dentibus (sub)aequalibus, usque ad circa tertiam partem divisi, plus minusve contractis in fructu; fauces pilosae. Corollae bilabiatae, usque ad circa tertiam partem incisae, calycibus circa 2plo longiores. Stamina inaequalia, plus minusve stricta, superiora inclusa, inferiora breviter exserta, filamentis corollis circa 2plo brevioribus.

Branches of the first order always present, those of the second order often so, and those of the third order sometimes present. *Leaves* herbaceous or somewhat leathery. *Spikes* (very) loose and tenuous, often each consisting of a whole branch, distinct from the rest of the plants, (very) long, erect. *Bracts* different from the leaves in shape and size, but not in texture and colour, slightly or not imbricate, $\frac{3}{4}$ – $1\frac{1}{3}$ x calyces. *Flowers* 2 per verticillaster, bisexual or female, (very) small. *Calyces* more or less tubular with 5 (nearly) equal teeth for c. $\frac{1}{3}$, when fruiting, teeth more or less contracted; throats pilose. *Corollas* 2-lipped for c. $\frac{1}{3}$, c. 2 x calyces. *Stamens* unequal in length, more or less straight, the upper 2 included, the lower 2 shortly protruding; filaments c. $\frac{1}{2}$ x corollas, glabrous.

31. Origanum elongatum (Bonnet) Emberger et Maire – Figs. 25 and 26.

O. elongatum (Bonnet) Emberger et Maire, Mem. Soc. Nat. Maroc 17: 43 (1927); Jahandiez & Maire, Cat. Pl. Maroc 3: 650 (1934). *O. glandulosum* Defontaines var. *elongatum* Bonnet, in de Segonzac, Voyages Maroc: 363 (1903). – Type: *de Segonzac s.n.*, Morocco, Djebel Beni Aziz (holo. P).

Woody perennials. Stems erect, up to 90 cm long, light or dark (purplish) brown, at the bases hirsute (hairs c. 1.5 mm long) otherwise glabrous, often glaucous. Branches of the first order present, in the upper $\frac{1}{3}$ – $\frac{1}{2}$ of the stems, up to 15 pairs per stem, 4 (1 – 20) cm long; branches of the second order sometimes present, those of the third order seldom so; all branches entirely or for the greater part consisting of spikes. *Leaves* up to 30 pairs per stem, shortly petiolate in the lower part to subsessile in the upper part (petioles up to 5 mm long), ovate or oval, margins entire, tops (±) obtuse, 10 (2 – 20) mm long, 8 (1 – 14) mm wide, somewhat leathery, light green or purplish, often glaucous, glabrescent (pilose to glabrous; hairs c. 1.2 mm long), sessile glands up to 1600 per cm^2. *Spikes* very loose and tenuous, 40 (10 – 140) mm long, c. 3 mm wide. *Bracts* 10 (3 – 22) pairs per spike, ± lanceolate, tops acute, 3 (2 – 4) mm long, 1 (0.7 – 1.5) mm wide, glabrous or pilosellous, green, often glaucous. *Flowers* (sub)sessile. *Calyces* 3.5 (3 – 4) mm long, outside glabrous or pilosellous; teeth 1 (0.8 – 1.3) mm long. *Corollas* 6 (5.5 – 6.5) mm long, pink, outside pilosellous; upper lips divided, for c. $\frac{1}{5}$, into 2, c. 0.3 mm long lobes; lower lips divided, for c. $\frac{3}{5}$, into 3, somewhat unequal, 1 (0.7 – 1.3) mm long lobes. *Staminal filaments* up to 2 and 3.5 mm long. *Styles* up to 8 mm long.

Geography and ecology. *O. elongatum* is found in northern Morocco in the Rif Mts. and the northern part of the Moyen Atlas, on schistose rocks from 400 – 1500 m. It flowers from June to October.
Note. *O. elongatum* is related to *O. floribundum* and *O. grosii*. From both it differs in the usually glabrous or glabrescent, glaucous stems and leaves, and also in the conspicuously glandular punctate, ± leathery leaves and very loose spikes.

MOROCCO: Djebel Beni Aziz, 3 Aug. 1901, *de Segonzac s.n.* (type). Near Tizzi Iffri (Rif Atlas), on stony hills on schist, c. 1500 m, 24 June 1927, *Font Quer 551* (G, M). Rif Atlas, Beni-Musa, on steep slopes, 1450 m, 5 July 1932, *Sennen & Maurico 8480* (G). Rif Atlas, Beni-Meydui, on schist, 19 June 1933, *Sennen & Maurico 8931* (G, JE). Rif Occidental, Talasentan, 17 Oct. 1957, *Ruiz de la Torre s.n.* (MA). Rif Occidental, Bab Bérret, 21 Oct. 1957, *Ruiz de la Torre s.n.* (MA).

32. **Origanum floribundum** Munby – **Figs. 26** and **27.**

O. floribundum Munby, Bull. Soc. Bot. Fr. 2: 286 (1855); Battandier & Trabut, Fl. Algerie 2: 675 (1884). – Type: *Duval-Jouve 1652*, Algeria, near Rovigo (holo. P).
O. cinereum de Noë, Bull. Soc. Bot. Fr. 2: 579 (1855). Type: *Jamin 199*, Algeria, Petit Atlas, ravin de l'Harrach (holo. P, iso. G, W).

Woody perennials. Stems erect or ascending, up to 60 cm long, (dark) brown or greyish, piloso-tomentose (hairs c. 1.5 mm long). Branches of the first order always present, in the upper $\frac{1}{5}$ – $\frac{3}{5}$ of the stems, up to 15 pairs per stem, 1 (0.5 – 15) cm long; branches of the second order often present, those of the third order sometimes so; all branches entirely or for the greater part consisting of spikes. *Leaves* up to 25 pairs per stem, shortly petiolate (petioles up to 12 mm long), heart-shaped, ovate or oval, tops obtuse or ± acute, 15 (3 – 23) mm long, 15 (2 – 25) mm wide, thin, greyish green, piloso-tomentose (hairs c. 1 mm long), sessile glands inconspicuous, up to 1700 per cm^2. *Spikes* ± loose and tenuous, 20 (4 – 90) mm long, c. 3 mm wide. *Bracts* 7 (2 – 23) pairs per spike, ± lanceolate, tops ± acute, 2.5 (1.8 – 4) mm long, 1 (0.7 – 2.0) mm wide, pilosello-tomentellous, greyish or purplish. *Flowers* (sub)sessile. *Calyces* 3 (2.4 – 4) mm long, outside pilosello-tomentellous; teeth 0.8 (0.5 – 1.5) mm long. *Corollas* 6.5 (4 – 7.5) mm long, pink, outside somewhat pilosellous; upper lips divided, for c. $\frac{1}{5}$, into 2, c. 0.4 mm long lobes; lower lips divided, for c. $\frac{1}{2}$, into 3, somewhat unequal, 1 (0.5 – 1.5) mm long lobes. *Staminal filaments* up to 2.5 and 4 mm long. *Styles* up to 9.5 mm long.

Geography and ecology. *O. floribundum* is found in mountain regions south of the city of Alger, from c. 300 – 1600 m. It flowers from July to November.
Note. *O. floribundum* differs from *O. elongatum* in its densely piloso-tomentose stems and leaves, its not leathery, not conspicuously glandular punctate leaves, and in its ± loose spikes. From *O. grosii* it is differing in its ± loose spikes and somewhat smaller bracts.

ALGERIA: Petit Atlas, gorge of l'Harrach, on rocks, 10 Aug. 1851, *Jamin 199* (P, G, W). Gorge of l'Harrach, 31 July 1853, *Durando s.n.* (G). Near Rovigo, 1853, *Duval-Jouve 1652* (type). Mountains of Ain-Selazit, near Blida, 10 July 1854, *Cosson s.n.* (JE, W). Gorge of l'Oued el Kebir near Blida, 13 July 1854, *Perreaudière s.n.* (G). Near Blida, 31 July 1861, *Lefebvre s.n.* (W). Near Blida, Aug. 1889, *Battandier & Trabut 563* (G, L). Blida, gorges of Chiffa, 20 Aug. 1918, *Cuénod s.n.* (G). S. of Blida, valley of the Qued Rebir, wooded places, c. 350 m, 4 Nov. 1929, *Zerny s.n.* (W). Atlas of Blida, clearings in cedar woods of Chréa, c. 1600 m, 25 July 1948, *Dubois & Faurel 1005* (G).

33. **Origanum grosii** Pau et Font Quer ex Ietswaart – **Figs. 25** and **26.**

O. grosii Pau et Font Quer ex Ietswaart (validated here). *O. grosii* Pau et Font Quer, Iter Maroccanum no. 352 (1928), *nomen nudum*; Jahandiez & Maire, Cat, Pl. Maroc 3: 650 (1934). – Type: *Font Quer 352*, Morocco, Mt. Kalaa (holo. MA, iso. G).

O. x *font-queri* Pau, Iter Maroccanum no. 578 (1930), *nomen nudum*; Jahandiez & Maire, Cat. Pl. Maroc 3: 650 (1934). – Type: *Font Quer 578*, Morocco, near Talambot (holo. MA, iso. G).

Species haec differt a *O. elongato* et *O. floribundo* spiciis brevioribus compactisque et bracteis amplioribus. Plantae perennes basibus lignosis. Caules erecti, usque ad 40 cm longi, basi pilosi, ceterum piloselli (pili 0.3 – 1.5 mm longi). Rami primarii semper adsunt, 2 (1 – 7) cm longi; rami secundarii interdum adsunt, rami tertiarii desunt. Folia inferiora petiolis usque ad 5 mm longis, superiora subsessilia, ovata vel elliptica, 8 (2 – 14) mm longa, 8 (1 – 12) mm lata, pilosella (pili 0.2 – 1 mm longi), margine integro, apice (plus minusve) obtuso, glandibus sessilis usque ad 1750 in cm^2. Spicae plus minusve compactae, 12 (4 – 30) mm longae, 4 (3 – 5) mm latae. Bracteae 6 (3 – 12) pares pro spica, ovatae usque ad plus minusve lanceolatae, 3.5 (2.5 – 5) mm longae, 2.5 (1 – 3) mm latae, pilosellae, (purpureo)-virides, apice acuto. Flores (sub)sessilia. Calyces 3 (2.5 – 3.5) mm longi, externe piloselli, dentibus 0.8 (0.7 – 1.1) mm longis. Corollae 6 (4 – 7.5) mm longae, roseae, externe pilosellae. Filamenta usque ad 2.5 et 4 mm longa. Styli usque ad 8 mm longi.

Woody perennials. Stems erect, up to 40 cm long, purplish brown, at the bases pilose, otherwise pilosellous (hairs 0.3 – 1.5 mm long). Branches of the first order present, in the upper $\frac{1}{3}$ – $\frac{1}{2}$ of the stems, up to 8 pairs per stem, 2 (1 – 7) cm long; branches of the second order often present, those of third order not so; all branches for the greater part consisting of spikes. *Leaves* up to 25 pairs per stem, shortly petiolate in the lower parts to subsessile in the upper parts (petioles up to 5 mm long), ovate or oval, margins entire, tops ($\pm$) obtuse, 8 (2 – 14) mm long, 8 (2 – 12) mm wide, light green, pilosellous (hairs 0.2 – 1 mm long), sessile glands up to 1750 per cm^2. *Spikes* $\pm$ compact, 12 (4 – 30) mm long, 4 (3 – 5) mm wide. *Bracts* 6 (3 – 12) pairs per spike, ovate to $\pm$ lanceolate, tops acute, 3.5 (2.5 – 5) mm long, 2.5 (1 – 3) mm wide, pilosellous, (purplish) green. *Flowers* (sub)sessile. *Calyces* 3 (2.5 – 3.5) mm long, outside pilosellous; teeth 0.8 (0.7 – 1.1) mm long. *Corollas* 6 (4 – 7.5) mm long, pink, outside pilosellous; upper lips divided, for c. $\frac{1}{5}$, into 2, c. 0.3 mm long lobes; lower lips divided, for c. $\frac{3}{5}$, into 3, somewhat unequal, 1 (0.5 – 1.3) mm long lobes. *Staminal filaments* up to 2.5 and 4 mm long. *Styles* up to 8 mm long.

Geography and ecology. *O. grosii* occurs very locally in northern Morocco, on calcareous slopes, from c. 650 – 1000 m. It has been found flowering in June and July.

Notes. 1. *O. grosii* as well as *O.* x *font-queri* originally have been described very shortly on the labels of the exsiccatae series Iter Maroccanum 1928 and 1930. So a

Latin diagnose had to be added for *O. grosii*. 2. Originally the epithet of the first species was spelled "crosii", while the authors later on spelled it "grosii". 3. The hybrid *O.* x *font-queri* (*O. grosii* x *compactum*?) does nearly not differ in the characters given by the author, nor in any other characters from *O. grosii*, so it is synonymized with this latter name. It is however quite possible that *O. grosii* and *O.* x *font queri* both must be considered as, slightly different, products of hybridization between *O. elongatum* and *O. compactum*. Further investigations are desirable. 4. From *O. elongatum* as well as from *O. floribundum*, *O. grosii* differs in the more compact spikes and somewhat larger bracts.

MOROCCO: Mt. Kalaa, on calcareous rocks, c. 1000 m, 29 June 1928, *Font Quer 352* (type). Near Talambot, calcareous slopes, c. 650 m, 8 July 1930, *Font Quer 577* (G). Ibid., *Font Quer 578* (G, MA).

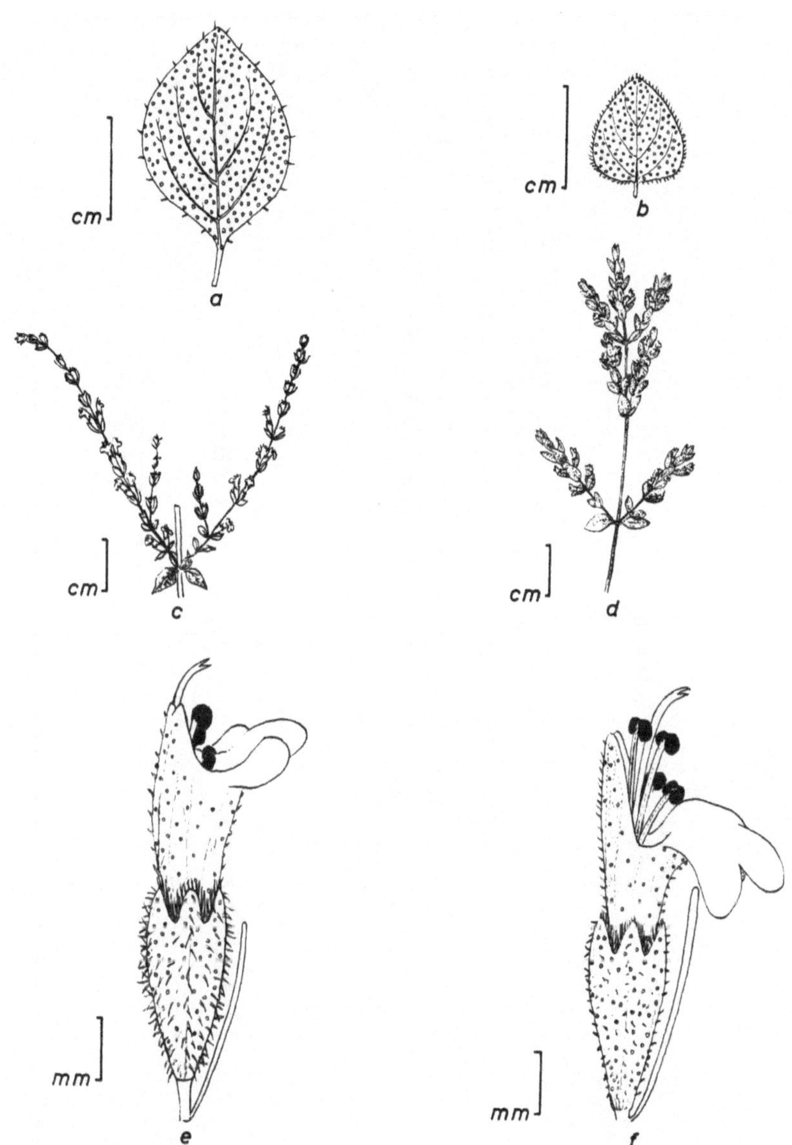

Figure 25. Leaves, spikes, and flowers with bracts in side view of the species in the section *Elongatispica*, except *O. floribundum* (for which see figure 27): a., c. and e. *O. elongatum;* b., d. and f. *O. grosii.*

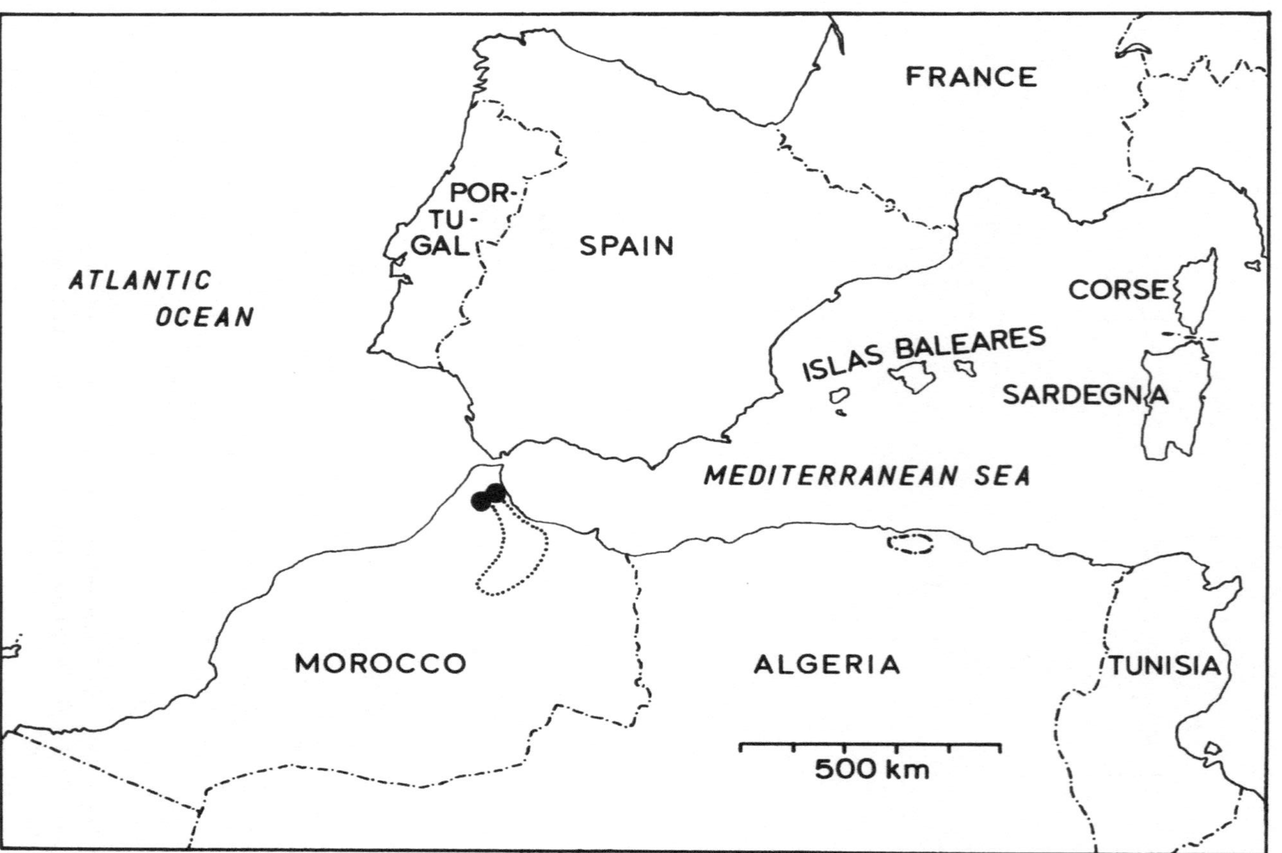

Figure 26. Distribution of the species in the section *Elongatispica*: .... *O. elongatum*; —·—·—. *O. floribundum*; ● *O. grosii*.

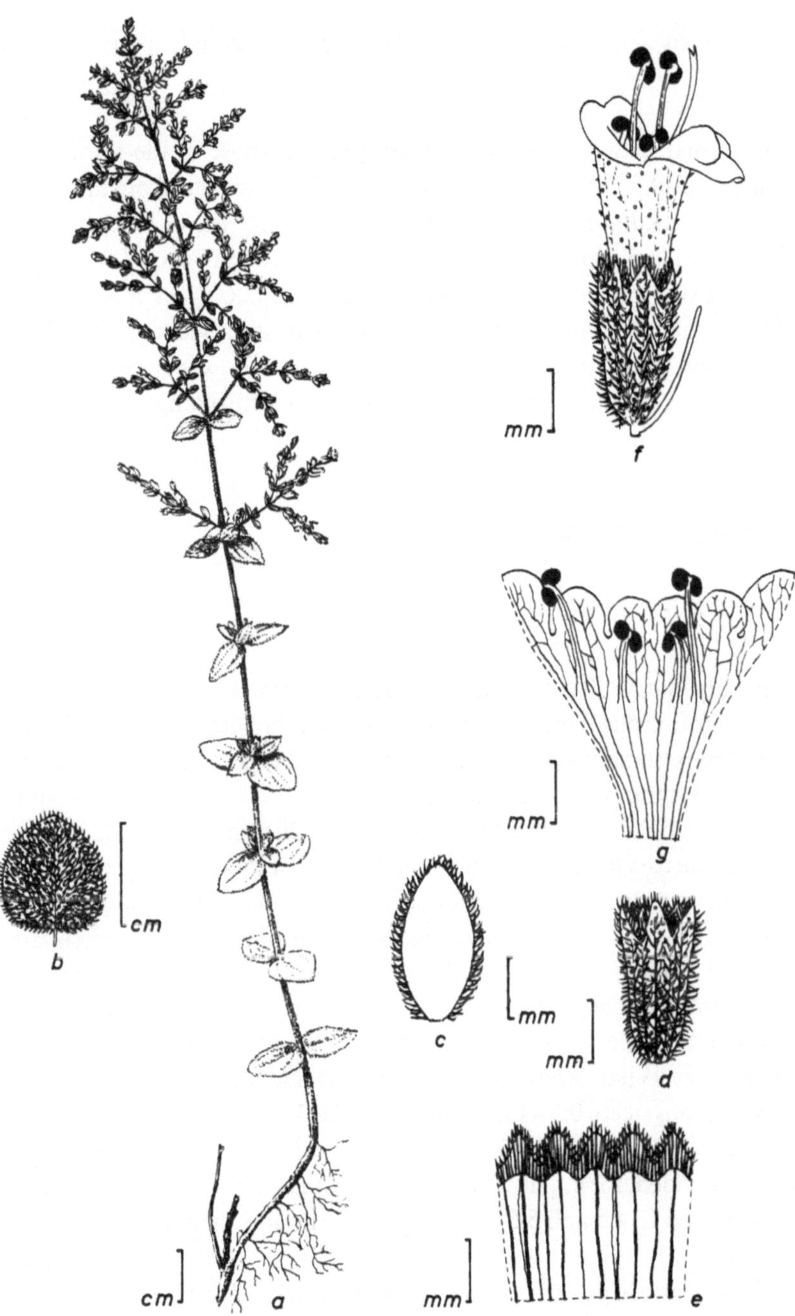

Figure 27. O. floribundum: a. habit; b. leaf; c. bract inside; d. calyx; e. calyx cut through the lower lip; f. flower with bract in side view; g. corolla cut through the lower lip.

IX. Section Origanum

Subgenus *Euoriganum* Vogel, Linnaea 15: 79 (1841). – Monotypic, type and only species: *Origanum vulgare* Linnaeus.

Branches of the first and second order always present, those of the third order usually so. *Leaves* usually herbaceous. *Spikes* (very) dense, distinct from the rest of the plants, small to medium-sized. *Bracts* different from the leaves in texture and colour and/or shape and size, (densely) imbricate, c. $\frac{1}{2}$ – 2 × calyces. *Flowers* 2 per verticillaster, bisexual or female, small or medium sized. *Calyces* more or less tubular with 5 (nearly) equal teeth for c. $\frac{1}{3}$, when fruiting teeth more or less contracted; throats pilose. *Corollas* 2-lipped for c. $\frac{1}{3}$, c. $2\frac{1}{2}$ × calyces. *Stamens* unequal in length, straight, subincluded or (shortly) protruding; filaments c. $\frac{1}{2}$ × corollas.

34. **Origanum vulgare** Linnaeus.

O. vulgare Linnaeus, Sp. Pl.: 590 (1753), (enlarged here). All synonyms are given under the subspecies. – Type: *Linnaeus 743.9* (holo. LINN).

Woody perennials. Stems 20 – 100 cm long, usually ascending and rooting at the bases, light to dark (purplish) brown, (appressed) pilose, hirsute (hairs 0.1 – 2.5 mm long), or ± glabrous, sometimes glaucous. Branches of the first order present, in the upper $\frac{1}{10}$ – $\frac{1}{2}$ of the stems, up to 12 pairs per stem, 0.5 – 25 cm long. *Leaves* up to 45 pairs per stem, petiolate to subsessile (petioles up to 20 mm long), ovate, oval or roundish, tops acute to obtuse, 6 – 40 mm long, 5 – 30 mm wide, hirsute or pilose to glabrous (hairs 0.1 – 2 mm long), sometimes glaucous, sessile glands hardly visible to very conspicuous, 100 – 2000 per cm^2, margins entire or remotely serr(ul)ate. *Spikes* 3 – 35 mm long, 2 – 8 mm wide. *Bracts* 2 – 25 pairs per spike, (ob)ovate or oval, tops ± acute or acuminate, 2 – 11 mm long, 1 – 7 mm wide, hirtellous, (densely) pilosellous or glabrous, (partly) purple, green or yellowish green, sometimes glaucous. *Flowers* (sub)sessile. *Calyces* 2.5 – 4.5 mm long, outside hirtellous, pilosellous or glabrous; teeth 0.5 – 1 mm long. *Corollas* 3 – 11 mm long, purple, pink or white, outside pilosellous; upper lips divided, for c. $\frac{1}{5}$, into 2, 0.2 – 0.7 mm long lobes; lower lips divided, for c. $\frac{1}{2}$, into somewhat unequal, 0.5 – 1.7 mm long lobes. *Staminal filaments* up to 4.5 and 5.5 mm long. *Styles* up to 13 mm long.

Geography and ecology. *O. vulgare* is an extremely variable species which ranges from the Azores, Madeira and the Canary Islands and Europe through the Mediterranean area, West and Central Asia to East Asia and Taiwan. It occurs from sea-level up to 4000 m. Mostly it is found on calcareous substrates, less frequently on non-limy soils. It flowers from May to October.

Note. Many names, on the species level and below have been given in the past to all kinds of morphological variation in the *O. vulgare* complex, e.g. with respect to degree of branching, indumentum, length of spikes, size and colour of bracts and size of flowers. A comprehensive study of many specimens resulted in the conclusion that indeed many species, subspecies and varieties can be discerned in their typical form, but that nearly all of them gradually pass into at least one other. So six rather variable subspecies are discerned which are not split up further into varieties. The subspecies are mainly based on differences in indumentum, number of sessile glands on leaves, bracts and calyces, and in size and colour of bracts and flowers.

a. ssp. **vulgare** – **Figs. 28** and **29.**

O. vulgare Linnaeus. *O. vulgare* Linnaeus ssp. *genuinum* Gaudin, Fl. Helv. 4: 77 (1829). *O. vulgare* Linnaeus emend. Koch, Linnaea 21: 661 (1848). *Thymus origanum* (Linnaeus) Kuntze, Taschenfl. Leipzig: 106 (1867). *O. vulgare* Linnaeus ssp. *euvulgare* Hayek, Prodr. Fl. Penins. Balc. 2: 334 (1931). *O. vulgare* Linnaeus ssp. *vulgare* var. *vulgare* f. *vulgare* Soo et Borhidi, Ann. Univ. Sc. Budap. 9 – 10: 361 (1968). *O. vulgare* Linnaeus var. *typicum* Fiori, Nuova Fl. Anal. Ital. 2: 455 (1969). Sibthorp & Smith, Prodr. Fl. Graec.: 418 (1809); Gray, Nat. Arr. Brit. Pl. 2: 380 (1821); Röhling & Koch, Deutschl. Fl. 4: 303 (1833); Savi, Osserv. Gen. Origanum: 8 (1840); Willkomm & Lange, Prodr. Fl. Hisp. 2: 398 (1868); Boissier, Fl. Or. 4: 551 (1879); Nyman, Consp. Fl. Eur.: 592 (1881); Battandier & Trabut, Fl. Algerie 2: 675 (1884); Hooker, Fl. Brit. India 4: 648 (1885); Briquet, in Engler & Prantl, Nat. Pflanzenfam. 4(3a): 309 (1895); Briquet, Lab. Alpes Marit. 3: 480 (1895); Halácsy, Consp. Fl. Graec. 2: 554 (1902); Fritsch, Exkursionsfl. Österr.: 451 (1922); Post & Dinsmore, Fl. Syr. Palest. Sin. 2: 334 (1933); Coutinho, Fl. Port.: 611 (1939); Rechinger, Fl. Aegaea: 531 (1943); Thiebaut, Fl. Libano-Syr. 3: 46 (1953); Wolf, Baileya 2: 62 (1954); Komarov, Fl. U.S.S.R. 21: 464 (1954); Briquet, Prodr. Fl. Corse 3: 216 (1955); Clapham et al., Fl. Brit. Isles: 739 (1962); Lid, Norsk Svensk Fl.: 584 (1963); Hedge & Lamond, Not. R.B.G. Edinburgh 28: 123 (1968); Polunin, Flow. Eur.: 364 (1969); Garcke, Ill. Fl. Deutschl.: 1225 (1972); Tutin et al., Fl. Eur. 3: 171 (1972); Hess et al., Fl. Schweiz 3: 145 (1972); Langhe et al., Nouv. Fl. Belgique: 416 (1973); Täckholm, Stud. Fl. Egypt: 458 (1974); Heukels & van Ooststroom, Fl. Nederl.: 536 (1977); Fournier, Quatre Fl. France: 839 (1977).

O. creticum Linnaeus, Sp. Pl.: 589 (1753). *O. vulgare* Linnaeus var. *creticum* (Linnaeus) Briquet, Lab. Alpes Marit. 3: 485 (1895). – Type: *Linnaeus 743.3* (holo. LINN).

O. majus Garsault, Traité Pl. Anim. 3: 256 (1767).

O. latifolium Miller, Gard. Dict. VIII Ed.: no. 3 (1768). – Type: *Miller s.n.* (holo. BM).

O. orientale Miller, Gard. Dict. VIII Ed.: no. 5 (1768). – Type: *Miller s.n.* holo. BM).

O. anglicum Hill, Veg. Syst. 17: 35 (1770).

O. purpurascens Gilibert, Indag. Nat. Lith.: 74 (1781).

O. floridum Salisbury, Prodr. Stirp.: 85 (1796).

O. vulgare Linnaeus var. *purpureum* Stokes, Bot. Mat. Med. 3: 345 (1812).

O. vulgare Linnaeus var. *rufuscens* Stokes, Bot. Mat. Med. 3: 346 (1812).

O. vulgare Linnaeus ssp. *prismaticum* Gaudin, Fl. Helv. 4: 78 (1829). *O. vulgare* Linnaeus var. *prismaticum* (Gaudin) Bentham, Lab. Gen. Sp.: 335 (1834).

O. vulgare Linnaeus ssp. *prismaticum* Gaudin var. *parviflorum* Gaudin, Fl. Helv. 4: 78 (1829).

O. vulgare Linnaeus ssp. *prismaticum* Gaudin var. *australe* Gaudin, Fl. Helv. 4: 78 (1829).

O. stoloniferum Besser ex Reichenbach, Fl. Germ. Exc.: 313 (1831).

O. thymiflorum Reichenbach, Fl. Germ. Exc.: 313 (1831).

O. decipiens Wallroth ex Bentham, Lab. Gen. Sp.: 728 (1834).

O. americanum Rafinesque, Fl. Tell. 3: 86 (1836). *O. vulgare* Linnaeus var. *americanum* (Rafinesque) Rafinesque, Aut. Bot. 1: 119 (1840). – Type: *Rafinesque s.n.*, U.S.A. (holo. G).

O. capitatum Willdenow ex Bentham, Linnaea 11: 339 (1837). – Type: *Willdenow s.n.* (holo. B).

O. nutans Willdenow ex Bentham, Linnaea 11: 339 (1837).

O. venosum Willdenow ex Bentham, Linnaea 11: 339 (1837). – Type: *Willdenow s.n.* (holo. B).

O. vulgare Linnaeus var. *rotundifolium* Rafinesque, Aut. Bot. 1: 119 (1840).

O. vulgare Linnaeus emend. Koch var. *spicatum* Koch, Linnaea 21: 661 (1848).
O. vulgare Linnaeus var. *pallescens* Martrin-Donos, Fl. Tarn: 550 (1864).
O. watsoni Schmidt et Schlagintweit, J. Bot. 6: 234 (1868).
O. vulgare Linnaeus var. *subglabrum* Schmidt et Schlagintweit, J. Bot. 6: 234 (1868).
O. vulgare Linnaeus var. *exile* Lamotte, Prodr. Fl. Plat. Centr. 2: 595 (1881).
O. barcense Simonkai, Természetrajzi Füzetek 10: 182 (1886). *O. vulgare* Linnaeus var. *barcense* (Simonkai) Hayek, Fl. Penins. Balc. 2: 334 (1931). *O. vulgare* ssp. *barcense* (Simonkai) Jávorka emend. Borhidi, Ann. Univ. Sc. Budap. 9 – 10, 361 (1968). – Type: *Simonkai s.n.*, Transylvania, Mt. Czenk (holo. BP).
O. vulgare Linnaeus f. *glabrescens* Beck, Fl. Hernstein: 244 (1884). – Type: *Beck s.n.*, Austria, Schneeberg (holo. PRC).
O. vulgare Linnaeus var. *latebracteatum* Beck, Ann. K.K. Naturh. Hofm. 2: 142 (1887). – Type: *Beck s.n.*, Yugoslavia (Hercegovina), near Konjica (holo. PRC).
O. vulgare Linnaeus f. *elongatum* Formánek, Österr. Bot. Zeitschr. 40: 92 (1890).
O. vulgare Linnaeus f. *albiflora* Šehovci ex Formánek, Österr. Bot. Zeitschr. 40: 92 (1890).
O. vulgare Linnaeus var. *puberulum* Beck, Fl. Nieder-Österr. 2: 993 (1893). *O. puberulum* (Beck) Klokov, in Kotov & Barbarich, Fl. RSS Ucr. 9: 290 (1960).
O. vulgare Linnaeus var. *purpurascens* Briquet, Lab. Alpes Marit. 3: 486 (1895).
O. vulgare Linnaeus var. *spiculigerum* Briquet, Lab. Alpes Marit. 3: 486 (1895). *O. humile* Miller var. *spiculigerum* (Briquet) Rouy, Fl. France 11: 348 (1909).
O. barcense Simonkay var. *macrostachyum* Grecescu, Consp. Fl. Roman.: 459 (1898). *O. vulgare* Linnaeus f. *grecescui* Soó, Acta Bot. Acad. Sc. Hung. 11: 249 (1965). *O. vulgare* Linnaeus ssp. *barcense* (Simonkay) Jávorka emend. Borhidi f. *gresescui* Soó, in Soó & Borhidi, Ann. Univ. Sc. Budap. 9 – 10: 361 (1968).
O. vulgare Linnaeus var. *formosanum* Hayata, Ic. Pl. Formos. 8: 102 (1919).
O. majoranoides hort. ex Gams, in Hegi, Ill. Fl. Mittel-Eur. 5: 2330 (1927).
O. elegans Sennen, Bol. Soc. Ibérica 32: 75 (1934).
O. vulgare Linnaeus var. *violacea* Sennen, Bol. Soc. Ibérica 32: 75 (1934).
O. vulgare Linnaeus var. *tauricum* Borissova, in Komarov, Fl. U.S.S.R. 21: 465 (1954).
O. vulgare Linnaeus var. *bracteosum* Petermann ex Soó et Borhidi, Ann. Univ. Sc. Budap. 9 – 10: 361 (1968).
O. vulgare Linnaeus f. *procumbens* Jakucs ex Soó et Borhidi, Ann. Univ. Sc. Budap. 9 – 10: 361 (1968).
O. vulgare Linnaeus ssp. *barcense* (Simonkay) Jávorka emend. Borhidi f. chlorescens Simonkay ex Soó et Borhidi, Ann. Univ. Sc. Budap. 9 – 10: 361 (1968).
O. vulgare Linnaeus ssp. *barcense* (Simonkay) Jávorka emend. Borhidi f. *pilosiusculum* Borhidi, in Soó & Borhidi, Ann. Univ. Sc. Budap. 9 – 10: 361 (1968).

Stems erect or ascending and rooting at the bases, up to 100 cm long, usually pilose, sometimes pilosellous or glabrescent (hairs c. 1 mm long). Branches up to 10 pairs per stem, 3.5 (0.2 – 16) cm long. *Leaves* up to 30 pairs per stem, usually ovate, tops obtuse to acute, 25 (3 – 50) mm long, 13 (2 – 33) mm wide, pilose(llous) or glabrescent (hairs c. 0.7 mm long), sessile glands usually not conspicuous, up to 800 per cm^2; margins remotely serr(ul)ate or entire; petioles up to 15 mm long. *Spikes* ovoid to cylindrical, 7 (3 – 35) mm long, 4 (3 – 7) mm wide. *Bracts* 6 (2 – 26) pairs per spike, (ob)ovate or oval, 4 (2 – 7) mm long, 2 (1 – 4) mm wide, usually membranous, glabrous or glabrescent, sometimes pilosellous, more or less vividly purple, sometimes glaucous. *Calyces* 3 (2 – 4) mm long. *Corollas* 7 (4 – 10) mm long, pink or purple. *Staminal filaments* up to 4 and 5 mm long. *Chromosome number* 2n = 30, 32.

Geography and ecology. *O. vulgare* ssp. *vulgare* is found all over the northern part of the distribution area of the species, i.e. from England and Scandinavia through Europe to Asia and Taiwan. From the rather scanty herbarium material from East Asia must be concluded that the ssp. does not occur frequently there. The occurrence in North America, at least since Linnaeus' time, is due to introduction by man. The ssp. *vulgare* is found from sea level up to 4000 m, in nearly the same habitats as ssp. *viride*. It flowers from June to November.

Notes. 1. Ssp. *vulgare* changes gradually into ssp. *viride* and ssp. *virens*. From both it differs in the clearly purple bracts and purple or pink corollas. 2. From *O. compactum*, ssp. *vulgare* differs in its smaller flowers and in its less glandular punctate leaves. 3. As in the ssp. *viride* forms occur with very short stems (c. 20 cm long). In both cases these have not been named separately. 4. Possibly one or two other subspecies may be discerned within *O. vulgare*, when more specimens become available from Central and East Asia. 5. With *O. majorana* ssp. *vulgare* formed hybrids (see p. 134).

NORWAY: Eide, 1897, *Greshoff s.n.* (L).

SWEDEN: between Norrskog and Levarnsvik, 14 Aug. 1929, *Asplund 1360* (L). Gotland, 6 Aug. 1930, *Fries s.n.* (W). Göteborg, Mölndal, Toltorpsdalen, 24 Sept. 1933, *Borgvall s.n.* (L). Islands near Fiskebäckskill, 17 Sept. 1961, *Duyfjes & Kanis 394* (L).

FINLAND: Nylandia, 18 July 1906, *Palmen s.n.* (W).

DENMARK: Skaelland, Jonstrup Vang, 14 Aug. 1885, *Mortensen s.n.* (L). Skaering, n. of Aarhus, 18 Sept. 1964, *Laegaard & Pedersen s.n.* (W).

ENGLAND: Caernavon, Llandudno, Aug. 1869, *Bailey 811* (W). Denbigh, s. of Llangollen, 22 Aug. 1877, *Bailey 961* (L). Surrey, 10 Aug. 1882, *Buysman 119* (L). Kent, the Warren, Folkestone, 3 Oct. 1906, *Bailey 1233B* (L). Dunnichen Hill, 17 Aug. 1913, *Corstorphine 1326* (L). Gloucester, Symonds Gat., 13 Aug. 1933, *Sandwith 1940* (L). Kent, Kemsing near Sevenoaks, Aug. 1951, *Hoogeveen s.n.* (L). Surrey, Box Hill, 23 July 1952, *Melderis & Bangerter 129* (L). Near Bath, 30 Aug. 1952, *Boom s.n.* (L).

NETHERLANDS: near Gulpen, Aug. 1917, *Dorgelo s.n.* (AVU). Near Veghel, 15 July 1939, *Prins s.n.* (AVU). Near Wageningen, 5 Aug. 1951, *Huizing s.n.* (AVU). Between Sippenaken and Beusdal, 22 July 1957, *Wattel s.n.* (AVU).

BELGIUM: near Namur, 15 Aug. 1868, *Thielens & Devos 320* (L). Falmignoul near Namur, 24 July 1878, *Gravet s.n.* (L). Citadel of Namur, 21 June 1924, *van Steenis s.n.* (L). Rémouchamps, June 1950, *van Borssum Waalkes 5147* (L).

LUXEMBOURG: s. of Grundhof, 6 July 1947, *Hoogland 1947 – 269* (L).

FRANCE: Gard, near Vigan, 10 Sept. 1862, *Billot 3451* (JE). Near Arnas, Aug. 1898, *Gandoger s.n.* (W). Auvergne, La Bourboule, Sept. 1923, *Kruijtbos-Heijnis s.n.* (L). Esquièze near Luz, 5 July 1925, *Ronniger s.n.* (W). Côte Vermeille, 20 July 1926, *Ronniger s.n.* (W). Dordogne, St. Vincent de Cosse, Sept. 1927, *van Soest s.n.* (L). Pyrénées Orientales, valley of Tet, near Mont Louis, 23 July 1944, *Rechinger & Sleumer 1212* (L). Haute Savoi, Mt. Salève, 24 Aug. 1946, *Paldeieux s.n.* (L). Near Grenoble, 19 July 1948, *Wissink 22* (L). Haut-Rhin, Zinnköpfle, w. of Rouffach, 1 July 1951, *Jacobs 811* (L).

SPAIN: Granada, near Langeron, 12 June 1873, *Winkler s.n.* (JE). Valencia, Férica, 4 Sept. 1911, *Pau 1363* (W). Catalona, Ripoll, Sept. 1913, *Sennen 2046* (W). Barcelona, massif of Tibidabo, July 1918, *Sennen 3535 & 3536* (W). Barcelona, Manileu, 9 Aug. 1927, *Gonzalo s.n.* (W). Santander, near Saja, 26 Aug. 1944, *Nartus & Vicioso s.n.* (MA). Bocequillas, s. of Burgos, 3 Aug. 1952, *de Wit 5273* (L). Altamira, 4 Aug. 1952, *de Wit 5273* (L).

B.R.D.: Oberlahnstein, Aug. 1861, *Wirtgen 706* (L). Eifel, Gerolstein, 12 Aug. 1928, *Koopmans-Forstmann & Koopmans s.n.* (L).

D.D.R.: Forst near Jena, 20 July 1901, *Popta s.n.* (L).

SWITZERLAND: near Ollon, 31 July 1926, *van Regteren Altena 1990* (L). Between Gsteig and Zweilütschinen, 1929, *Bayer 90* (L). Vaud, near Buchillon, Aug. 1952, *van Soest s.n.* (L).

AUSTRIA: Schneeberg, Aug. 1886, *Beck s.n.* (W). Sankt Johann, 31 July 1936, *Brand s.n.* (L). Burgenland, Leithagebirge, near Müllendorf, 23 Oct. 1965, *Melzer s.n.* (W).

ITALY: Liguria, above Sestri Ponente, 5 Sept. 1892, *Haussknecht s.n.* (JE). Novara, Arona, 12 Sept.

1893, *St.-Lager s.n.* (L). Bergamo, between Vilminore and Schilpario, 23 Sept. 1906, *St.-Lager s.n.* (L). Valle Formazza, between St. Antonio di Vova and Altillone, 10 Sept. 1912, *Boggiani s.n.* (L, W).

YUGOSLAVIA: near Orahovac, 25 July 1886, *de Szyszylowicz s.n.* (W). Mt. Dziebere, 31 July 1886, *de Szyszylowicz s.n.* (W). Dalmatia, above Cattaro, 1908, *Zahlbruckner s.n.* (W). Near Nevesinje, 16 Aug. 1911, *Schneider s.n.* (W). Istria, near Semich, 23 Aug. 1925, *Cufodontis s.n.* (W).

POLAND: Hultschin, Weinberg, 18 July 1892, *Ziesche s.n.* (L). Bielany near Krakow, 22 Nov. 1912, *Raciborski s.n.* (L). Pieniny, 17 Aug. 1938, *Walas 358* (L). Ciekowice, near Tarnow, 11 Aug. 1952, *Tacik 571* (L).

CZECHOSLOVAKIA: near Bratislava, 12 Aug. 1910, *Korb s.n.* (W). Near Borac, Aug. 1924, *Suza s.n.* (W).

RUMANIA: Corona, Mt. Zinne, 12 July 1893, *Sagorski s.n.* (JE). Ibid., 12 July 1895, *Sagorski s.n.* (JE). Mt. Czenk, near Brasso, 19 Aug. 1906, *Zsak s.n.* (JE).

BULGARIA: near Sliven, 16 July 1907, *Schneider 441* (W). Abraszow, Tschiflik near Russe, 13 July 1930, *Ronniger s.n.* (W). Mt. Vitosa, near refuge Tintjava, 1 Aug. 1951, *Efremov s.n.* (W).

TURKEY: PROV. KIRKLARELI: near Dereköy, 24 July 1968, *Baytop 14138A* (E). PROV. TEKIRDAĞ: e. of Tekirdağ, 13 June 1968, *Baytop 13360* (E). PROV. ISTANBUL: Floria, San Stefano, 2 Aug. 1896, *Aznavour s.n.* (G). Kartal, Yakadjik, 29 July 1900, *Aznavour s.n.* (G). Bourounsouz Mandra, 13 July 1918, *Aznavour s.n.* (G). PROV. KASTAMONU: s. of Kastamonu, 3 July 1969, *Sorger 69 – 13 – 59* (Herb. Sorger). PROV. ÇANKĬRĬ: s. of Çankĭrĭ, 21 Aug. 1971, *Buttler 15567* (Herb. Buttler). PROV. RIZE: Ikizdere, Yetimhoca, 1952, *Davis 20990* (BM). PROV. KARS: Yalnizçam, 19 Aug. 1957, *Davis 32497* (E). Between Karakurt and Sarikamis, 14 July 1966, *Davis 46579* (E).

U.S.S.R.: Caucasus, near Elisabethpol, 1834, *Hohenhacker s.n.* (G, W). Mosqua, near Uspenskoje, 6 Aug. 1900, *Choroschkov 3785A* (W). Saratow, near Petrowsk, Aug. 1900, *Blumberg 986* (W). Ucraina, near Kiev, 10 Sept. 1907, *Dumansky 3785C* (W). Tomsk, Zmeïnogorsk, 27 June 1909, *Iljin s.n.* (L). Krasnojarsk, Enisseisk, 8 July 1914, *Kuznetzov s.n.* (W). Caucasus, Georgia, Tiblisi, Dabahane gorge, 30 June 1959, *Davis 33886* (E).

IRAN: s. of Chalus, 27 June 1962, *Furse 2877* (W). Mazanderan, 14 June 1966, *Archibald 2321* (E). Azerbaijan, Kaleybar, 19 July 1971, *Lamond 4837* (E).

INDIA: Northwest Himalaya, *Hooker & Thomson s.n.* (L).

CHINA: East Thibet, Ta-tsien-lou, 1893, *Soulie s.n.* (G). Tong-tchouan, Oct. 1912, *Maire s.n.* (G). Yunnan, June 1908, *d'Alleizette s.n.* (L). Northwest Setschwan, near Sungpan, 1914, *Weigold s.n.* (W). Yunnan, Mekong-Yangtze, Sept. 1921, *Forrest 20372* (W). Sze-chu'an, Nsü-tsing, *Smith 4785* (W).

b. ssp. **glandulosum** (Desfontaines) Ietswaart – **Figs. 29, 30** and **31.**

O. vulgare Linnaeus ssp. *glandulosum* (Desfontaines) Ietswaart *stat. nov. O. glandulosum* Desfontaines, Fl. Atl. 2: 27 (1799); Boissier, Fl. Or. 4: 552 (1879); Battandier & Trabut, Fl. Algerie 2: 675 (1884); Bonnet & Barratte, Cat. Pl. Vasc. Tunesie: 328 (1896). – Type: *Desfontaines s.n.*, Algeria, near Mascara (holo. P).

Stems ± erect, up to 100 cm long, hirsute (hairs c. 1.5 mm long). Branches up to 18 pairs per stem, 3.5 (0.2 – 25) cm long. *Leaves* up to 35 pairs per stem, (broadly) ovate or oval, tops obtuse or acute, 18 (4 – 38) mm long, 13 (2 – 27) mm wide, hirsute (hairs c. 1 mm long), sessile glands very conspicuous, up to 2000 per cm^2; margins entire or remotely serrulate; petioles up to 15 mm. *Spikes* ovoid, seldom cylindrical, 5 (3 – 17) mm long, 4 (3 – 5) mm wide. *Bracts* 5 (2 – 15) pairs per spike, ovate to lanceolate, 3 (2 – 4) mm long, 1.2 (1 – 2) mm wide, herbaceous, glabrous or slightly pilosellous along the margins, green. *Calyces* 3.5 (3 – 4) mm long. *Corollas* 7 (5 – 8) mm long, white. *Staminal filaments* up to 3.5 and 5 mm long.

Geography and ecology. *O. vulgare* ssp. *glandulosum* is found all over the northern part of Algeria, and in Tunesia. It occurs in mountain regions, up to c.

1200 m, usually in rocky places. It has been found flowering from May to August. Note. Ssp. *glandulosum* is most closely related to ssp. *hirtum*, from which it differs in the (±) glabrous, clearly glandular punctate bracts, which are usually somewhat shorter than the calyces. From the partly sympatric *O. floribundum* it differs in the hirsute stems and leaves, and the compact spikes.

ALGERIA: hills near Kouba, May 1837, *Bové s.n.* (G, L). Near brooklet Qued-Smar, 2 Aug. 1848, *Salle 67* (JE). Mts. of Kouba near Alger, June 1850, *Jamin 74* (G, W). Near Oran, May 1851, *Munby 69* (P). Near Alger, 1853, *Durando s.n.* (G). Maison Carré near Alger, 24 June 1854, *Durando 19* (G, W). Mts. of Ouarsenis, 25 July 1854, *Cosson s.n.* (JE, W). S. of Tlemcen, 18 June 1856, *Bourgeau 39* (G, P, W). Sétif near Constantine, 24 June 1856, *Choulette 79* (W). Nitidja near Blida, 19 June 1862, *Lefebvre s.n.* (W). Mt. Sidi Mecid near Constantine, 22 June 1869, *Paris 370* (JE). Constantine, May 1877, *Reboud 1791* (G). Le Corso, June 1886, *Battandier & Trabut 364* (G, L). Djbel Ouach near Constantine, June 1888, *Girod s.n.* (G). Kerata, July 1896, *Reverchon 77* (G, JE). Tlemcen, 24 June 1906, *Faure s.n.* (G). El Biar near Alger, 27 June 1916, *Bianor 2900* (G, W). Tlemcen, 16 July 1928, *Briquet 1466* (G). Tlemcen, 26 Aug. 1932, *Faure s.n.* (G).

TUNESIA: hills near Gabes, June 1854, *Kralik s.n.* (W). Djebel Zaihouan, 13 July 1854, *Kralik 290* (G, W).

c. ssp. **gracile** (Koch) Ietswaart – **Figs. 29, 30** and **31.**

O. vulgare Linnaeus ssp. *gracile* (Koch) Ietswaart, *stat. nov. O. gracile* Koch, Linnaea 21: 661 (1848). – Type: *Kotschy 444*, Turkey, near Musch (neo. G, JE, L, W).
O. bucharicum Bornmüller, Beih. Bot. Centralbl. 33: 308 (1915), *nomen nudum.*
O. tytthanthum Gontscharov, Delect. Sem. Inst. Bot. Sect. Tadshik.: 12 (1934).
O. tytthanthum Gontscharov var. *seravschianum* Borissova, in Komarov, Fl. URSS 21: 468 (1954).
O. kopetdaghense Borissova, Not. Syst. Herb. Inst. Bot. Nom. Komarovii 16: 280 (1954).
O. glaucum Rechinger et Edelberg, Dan. Biol. Skr. 8: 76 (1954). *O. vulgare* Linnaeus var. *glaucum* (Rechinger et Edelberg) Hedge et Lamond, Not. R. B. G. Edinburgh 28: 124 (1968). – Type: *Edelberg 1198*, Afghanistan, Nuristan, Chetras (holo. W, iso. C).
O. glaucum Rechinger et Edelberg var. *laxius* Rechinger et Edelberg, Dan. Biol. Skr. 8: 76 (1954). – Type: *Edelberg 1393*, Afghanistan, Nuristan, near Faizabad (holo. W, iso. C).

Stems ± erect, slender, up to 90 cm long, usually glabrescent, sometimes somewhat pilosellous (hairs c. 0.5 mm long), usually glaucous. Branches slender, up to 10 pairs per stem, 2.5 (0.5 – 18) cm long. *Leaves* up to 25 pairs per stem, (longly) ovate, tops usually ± acute, 18 (3 – 40) mm long, 11 (1 – 5 – 22) mm wide, glabrescent or slightly pilosellous (hairs nearly as long as those on the stems), usually glaucous, sessile glands usually conspicuous, 150 – 2000 per cm^2; margins entire or remotely serrulate; petioles up to 14 mm long. *Spikes* ovoid to cylindrical, 6 (4 – 30) mm long, 3 – 4 mm wide. *Bracts* 4 (2 – 26) pairs per spike, longly (ob)ovate or oval, 4 (3 – 6) mm long, 1.5 (1 – 2) mm wide, herbaceous or ± membranous, glabrous or pilosellous, green or slightly purple, usually glaucous. *Calyces* 3 (2 – 4) long. *Corollas* 6.5 (5 – 8) mm long, white or pink. *Staminal filaments* up to 2.5 and 3.5 mm long.

Geography and ecology. *O. vulgare* ssp. *gracile* is found in eastern Turkey, northern Iraq, northern Iran, northern Afghanistan and southern Central U.S.S.R. It occurs in mountain regions from 1000 – 3000 m, in various habitats: stony slopes (calcareous as well as non-limy substrates), gravel terraces and along streams, under dry as well as rather moist conditions. It flowers from June to September.

Note. Ssp. *gracile* touches ssp. *viride* and ssp. *vulgare*. From both it differs in at least two of the following characters: slender habit, glabrescent and glaucous stems and leaves, and clearly glandular punctate leaves. In the first characters it also differs from ssp. *hirtum*.

TURKEY: Mt. Avroman and Schahu, June-July 1867, *Haussknecht s.n.* (W). Between Hassanova and Kutschai, 14 July 1890, *Sintenis 2940* (JE, W). Mt. Deli Dagh, July 1893, *Bornmüller 3481* (E, W). Near Goulon, 15 Aug. 1906, *v/d Post s.n.* (G). Near Tschermisch, 9 July 1910, *Handel-Mazzetti 1936* (W). PROV. MARAS: spurs of Beryt Dagh, 13 Aug. 1865, *Haussknecht s.n.* (W). Distr. Cardak, 24 July 1952, *Davis 20237* (E). PROV. MALATYA: Dedeyazi-huyo, 13 Aug. 1966, *Peşmen 1040* (G). PROV. TUNCELI: Ovacik. 20 July 1957. *Davis 31441* (E). PROV. HAKKARI: Cilo Dağ, 6 Aug. 1954. *Davis 23932* (BM. E). PROV. BITLIS: near Bitlis, 10 Aug. 1910. *Handel-Mazzetti 2892* (W). C. 5 km n. of Bitlis, 5 July 1951, *Huber-Morath 11382* (Herb. Huber-Morath). Kamboe Dağ, above Hurmuz, 31 June 1954, *Davis 23447* (BM, E). Bitlis Gorge below Tutu, 16 Aug. 1956, *McNeill 603* (E). C. 4 km w. of Resadiye, 7 July 1966, *Davis 46051* (E). Tatvan, 26 July 1966, *Tong 13* (E). Ibid., *Tong 34* (E). PROV. VAN: Satak, Kavussahap Dağ, 24 July 1954, *Davis 23076* (BM, E). PROV. MUŞ: near Musch, 9 Sept. 1859, *Kotschy 444* (type).

IRAQ: Sikreen near Gara, Mt. Mosul Liwa, 26 July 1961, *Agnew 752* (E). Serzang, *Wheeler Haines 412* (E). Doroki near Serzang, 6 Aug. 1961, *Wheeler Haines s.n.* (E).

IRAN: Kurdestan, 4 Aug. 1967, *Iranshaki ü Termé s.n.* (W).

AFGHANISTAN: Khash Dt, 8 Aug. 1937, *Koelz 12926* (W). Nuristan, Chedras, 29 July 1948, *Edelberg 1198* (W). Faizabad, 1948, *Edelberg 1393* (W). Nuristan, Darim valley, from Darim upward, 6 Aug. 1951, *Neubauer 988* (W). Salang, s. side, 18 Sept. 1964, *Neubauer 4404* (W). Salang, n. side, 9 Oct. 1964, *Neubauer 5004* (W). Salang, Charikar, 25 June 1965, *Rechinger 31325* (W). Prov. Parvan, Salang Pass s. side, 80 km from Kabul, 25 June 1965, *Lamond 2065* (E). Salang, 1 July 1965, *Rechinger 31629* (W). Prov. Takhar, Khost-o-fereng, 18 July 1965, *Podlech 11881* (E). Prov. Laghman, Alishang, middle part of Darrah Rastyon, 16 July 1969, *Wendelbo 9719* (E, W). Prov. Baghlan, Khinjan valley, n. side of Salang Pass, 27 June 1969, *Wendelbo 9242* (E). Prov. Konar, Konar valley, below Capakok, 26 Aug. 1970, *Anders 5240* (W).

U.S.S.R.: Central Asia, Pamiro-alai, Mt. Babatag, 15 June 1936, *Gomolitzky & Federov 272* (W).

d. ssp. **hirtum** (Link) Ietswaart – **Figs. 29, 30** and **31.**

O. vulgare Linnaeus ssp. *hirtum* (Link) Ietswaart, *stat. nov.* *O. hirtum* Link, Enum. Pl. Horti Berol. 2: 114 (1822); Röhling & Koch, Deutschl. Fl. 4: 304 (1833); Raulin, Acta Soc. Linn. Bordeaux 24: 523 (1861); Boissier, Fl. Or. 4: 552 (1879); Battandier & Trabut, Fl. Algerie 2: 675 (1884); Lojacono Pojero, Fl. Sicula 2: 195 (1904); Holmes, Perf. Ess. Oil Rec. 4: 72 (1913). *O. hirtum* Link var. *genuinum* Vogel, Linnaea 15: 80 (1841). *O. vulgare* Linnaeus var. *hirtum* Visiani, Fl. Dalm. 2: 192 (1847). *O. hirtum* Link var. *typicum* Candargy, Bull. Soc. Bot. France 44: 459 (1897). – Type: *Sintenis & Bornmüller 848*, Greece, Makedhonia, Peninsula Hagion Oros (neo. G, W).

O. megastachyum Link, Enum. Pl. Horti Bot. Berol. 2: 114 (1822). *O. vulgare* Linnaeus var. *megastachya* (Link) Koch, in Röhling & Koch. Deutschl. Fl. 4: 304 (1833).

O. smyrnaeum sensu Sibthorp et Smith (non Linnaeus), Fl. Graeca 6: 57 (1826); Savi, Osserv. Gen. Origanum: 4 (1840). – Type: *Sibthorp s.n.* (holo. OXF).

O. creticum auct. non Linnaeus. *O. heracleoticum* auct. non Linnaeus var. *creticum* (auct. non Linnaeus) Halácsy, Consp. Fl. Graec. 2: 555 (1902); Quézel & Contandriopoulos, Natur. Monspel., Ser. Bot. 16: 137 (1965).

O. heracleoticum auct. non Linnaeus; Miller, Gard. Dict., VIII Ed.: no. 2 (1768); Reichenbach, Fl. Germ. Exc.: 313 (1831); Host, Fl. Austr. 2: 156 (1831); Koch, Linnaea 21: 662 (1848); Halácsy, Consp. Fl. Graec. 2: 555 (1902): Hayek, Prodr. Fl. Penins. Balc. 2: 334 (1931); Rechinger, Fl. Aegaea: 532 (1943); Wolf, Baileya 2: 62 (1954); Briquet, Prodr. Fl. Corse 3: 219 (1955); Rechinger, Bot. Jahrb. 80: 395 (1961); Tutin et al., Fl. Eur. 3: 171 (1972). *O. vulgare* Linnaeus ssp. *heracleoticum* (auct. non Linnaeus) Holmboe, Stud. Veg. Cyprus: 162 (1914).

O. hirtum Link var. *prismaticum* Vogel, Linnaea 15: 80 (1841).
O. illiricum Scheele, Flora, Neue Reihe 1: 574 (1843).
O. latifolium Scheele, Flora, Neue Reihe 1: 574 (1843).
O. hirtum Link var. *humile* Bentham, in de Candolle, Prodr. 12: 194 (1848).
O. hirtum Link f. *albiflorum* Haussknecht, Mitt. Thüring. Bot. Ver., Neue Folge 11: 49 (1897). *O. heracleoticum* auct. non Linnaeus var. *albiflorum* (Haussknecht) Halácsy, Consp. Fl. Graec. 2: 555 (1902).
O. hirtum Link f. *rubriflorum* Haussknecht, Mitt. Thüring. Bot. Ver., Neue Folge 11: 50 (1897). *O. heracleoticum* auct. non Linnaeus var. *rubriflorum* (Haussknecht) Halácsy, Consp. Fl. Graec. 2: 555 (1902).
O. hirtum Link f. *prismaticum* Haussknecht, Mitt. Thüring. Bot. Ver., Neue Folge 11: 50 (1897).
O. hirtum Link f. *trichocalycinum* Haussknecht, Mitt. Thüring. Bot. Ver., Neue Folge 11: 50 (1897). *O. heracleoticum* auct. non Linnaeus var. *trichocalycinum* (Haussknecht) Halácsy, Consp. Fl. Graec. 2: 555 (1902). *O. heracleoticum* auct. non Linnaeus f. *trichocalycinum* (Haussknecht) Rechinger, Ann. Naturh. Mus. Wien 43: 327 (1928).
O. hirtum Link var. *corymbulosum* Candargy, Bull. Soc. Bot. France 44: 459 (1897).
O. hirtum Link var. *laxiflorum* Candargy, Bull. Soc. Bot. France 44: 459 (1897).
O. hirtum Link var. *macrostachyum* Candargy f. *macrostachyoides* Candargy, Bull. Soc. Bot. France 44: 459 (1897).
O. hirtum Link var. *oostachyum* Candargy, Bull. Soc. Bot. France 44: 459 (1897).
O. hirtum Link var. *subtypicum* Candargy, Bull. Soc. Bot. France 44: 459 (1897).
O. heracleoticum auct. non Linnaeus var. *creticum* (auct. non Linnaeus) Halácsy f. *glabra* Halácsy, Consp. Fl. Graec. 2: 555 (1902).
O. heracleoticum auct. non Linnaeus var. *creticum* (auct. non Linnaeus) Halácsy f. *hirsuta* Halácsy, Consp. Fl. Graec. 2: 555 (1902).

Stems ± erect, up to 100 cm long, usually hirsute (hairs c. 1 mm long). Branches up to 10 pairs per stems, 1.5 (0.2 – 12) cm long. *Leaves* up to 35 pairs per stem, ovate or oval, tops obtuse to acute, 15 (2 – 33) mm long, 11 (1 – 30) mm wide, (densely) pilosellous (hairs c. 0.5 mm long), sessile glands conspicuous, up to 2000 per cm^2; margins entire or remotely serr(ul)ate; petioles up to 12 mm long. *Spikes* ovoid, sometimes cylindrical, 6 (3 – 35) mm long, 4 (3 – 6) mm wide. *Bracts* 5 (2 – 25) pairs per spike, (ob)ovate or oval, 3 (1.5 – 5) mm long, 1.5 (1 – 3) mm wide, (±) herbaceous, usually pilosellous, sometimes glabrescent, green, sometimes slightly purple. *Calyces* 2.5 (2 – 3.5) mm long. *Corollas* 6 (3 – 7.5) mm long, white, seldom slightly pink. *Staminal filaments* up to 4 and 5 mm long. *Chromosome number* 2n = 30.

Geography and ecology. *O. vulgare* ssp. *hirtum* is an East Mediterranean taxon, found mainly on the Balkan Peninsula and in Turkey, furtheron it is reported from Ischia and Malta. It grows from sea level to 1500 m, usually on dry sunny places (but sometimes in shade), on calcareous as well as on non-limy substrates, on grassy hills, abandoned fields, walls, and in maquis and pine woods. It has been found flowering from May to December.

Notes. 1. Study of the types revealed that *O. heracleoticum* must be considered as a synonym of *O. vulgare ssp. viride*, and *O. creticum* as a synonym of *O. vulgare* ssp. *vulgare*. 2. *O. vulgare* ssp. *hirtum* is related to ssp. *glandulosum*, from which it differs in the larger bracts, which are not or slightly glandular punctate, and the often shorter branches. From ssp. *viride* (with which it sometimes has been confused) and ssp. *vulgare* it differs in the abundantly glandular punctate leaves. 3. Most probably ssp. *hirtum* hybridizes in nature with at least five other *Origanum* species: *O. micranthum*, *O. microphyllum*, *O. onites*, *O. scabrum* and *O. sipyleum* (see pp. 141, 139, 136, 137 and 142).

YUGOSLAVIA: Istria, near Pola, 30 June 1855, *Mirich s.n.* (L). Ibid., 13 Sept. 1881, *Sintenis s.n.* (BM). Ibid., 27 Aug. 1874, *Freyn s.n.* (JE). Spalato, near Castelvecchio, June 1901, *Krebs s.n.* (JE). Ibid., June 1902, *Poscharsky s.n.* (COI). Island Lesina, 19 June 1910, *Keller s.n.* (W). Lissa, 7 July 1910, *Richter s.n.* (G, JE, W). Spalato, 17 June 1912, *Sagorsky s.n.* (JE). Doiran, near Kaluckova, 30 June 1917, *Bornmüller 1834* (JE). Lapad, near Gravosa (Gruz), 9 June 1923, *Zerny s.n.* (W). Spalato, 18 June 1927, *Korb s.n.* (W). Herzegowina, Sutorina, 23 July 1929, *Ronniger s.n.* (W). Near castle Stari n.w. of Split, 20 July 1939, *Oberneder 6633* (BM). Split, s. of Karjan, 12 July 1940, *Radermacher 431* (L). Split, n.w. of Karjan, 29 July 1940, *Radermacher 475* (L). Split, Bay of Zeuta, 8 Aug. 1940, *Radermacher 547* (L). Spalato, *Petter 266* (BM, W). Spalato, Mt. Biokovo, *Pichler & Handel-Mazzetti 181* (COI, G, JE, L, W).

ALBANIA: near Sarandë, 15 July 1933, *Alston & Sandwith 2236* (BM).

GREECE. MAINLAND: Peninsula Hagion Óros, Kerai, in woods, 25 June 1891, *Sintenis & Bornmüller 848* (type). Mt. Taygetos, above Koumousta, July 1844, *Heldreich s.n.* (L, W). Agrapha (Dolopia), Mt. Pindus, June 1885, *Heldreich s.n.* (W). Ibid., June 1885, *Haussknecht s.n.* (BM, JE). Makedhonia, Mt. Olympos, 29 July 1891, *Sintenis & Bornmüller 1404* (G). Mt. Pindus, near Sermeniko, 15 Aug. 1896, *Sintenis 1515* (G). Valley of Bouraïkos, near Zachlorou, 6 July 1896, *St. Lager s.n.* (G). Mt. Taygetos, 2 Aug. 1934, *Regel s.n.* (G). Reutino, Mt. Pindus, 5 Aug. 1937, *Balls & Balfour-Gourlay s.n.* (BM). Makedhonia, Kavalla, 20 Sept. 1957, *Rechinger 15788* (W). Peloponnesus, Laconia, Peninsula Malea, 8 June 1958, *Rechinger 20035* (W). Phtiotis, between Makrokomi and Rendina, 6 July 1958, *Rechinger 20606* (G, W). Thraki, Nomarchia Evros, near Esimi, July 1963, *Bauer 105* (W). Ibid., 1966, *Bauer & Spitzenberger 1153* (W). Makedhonia, Nomos Kavalas, Mt. Pangeon, 14 July 1970, *Strid 843* (W). CORFU: near Spartilla, 26 July 1936, *Regel s.n.* (G). Mt. Pantocrater, 6 Aug. 1968, *Bally 13162* (G). SAMOTHRACI: between Camariotissa and Samothraci, 30 June 1965, *Phitos 3196* (W). SKIATHOS: 20 Sept. 1966, *Phitos 5388* (W). EUBOEA: 1930, *Guiol s.n.* (BM). Mt. Delphi, July 1932, *Rechinger 2519* (BM). Mt. Delphi, near spring Liri, 26 July 1938, *Regel s.n.* (G). Mt. Ocha, below Hagios Dimitrios, May 1955, *Rechinger 16905* (W). Mt. Kandili, near Achmet Aga, 20 July 1956, *Rechinger 18181* (G, W). Mt. Kandili, between Psachna and Achmet Aga, 20 July 1956, *Rechinger 18194* (G, W). Near Kyni, June 1958, *Rechinger 18711* (W). Near Platana, 20 July 1965, *Phitos 4030* (W). ANDROS: near Arni, 19 July 1938, *Regel s.n.* (G). IKARIA: Mt. Atheras, 24 July 1887, *Forsyth Major 726* (G). NAXOS: Keramiti, July 1897, *Heldreich & Halácsy s.n.* (COI). July 1897, *Leonis s.n.* (G). Near Keramiti, 15 June 1898, *Leonis s.n.* (W). KRITI: la Canée, 7 July 1883, *Reverchon 134* (G, JE). Rethymo, Petrasnero, 29 May 1915, *Gandoger 12388* (G). Rethymo, Sybritos, 4 June 1915, *Gandoger 12668* (G). Canée, 11 June 1915, *Gandoger 7649* (G). Near Ennea Choria, June 1932, *Guiol 1393* (BM). Chania, between Skines and Nea Rumata, 29 May 1942, *Rechinger 13386* (BM, W). Selinos, near Vutas, 2 June 1942, *Rechinger 13554* (G, W). Chania, near Platanias, 17 July 1942, *Rechinger 14429* (BM, G, W). Levka Ori, near Omalos, 3 July 1943, *Rechinger 15048* (BM, G, W). Kidonia, 5 July 1961, *Greuter S3757* (G, W). Vassilios, near Sellià, 10 June 1962, *Greuter S4704* (G, W).

TURKEY: near Kareikos, 1883, *Sintenis 45613* (E). Jeniköi, Aug. 1900, *Schwöder s.n.* (W). Schato Dagh, 19 July 1903, *Aznavour s.n.* (G). SEYHAN: Bulgar Dağ, between Pozant and Meydan Yayla, 1 Sept. 1949, *Davis 16588* (E, G, W). S. end of Olukisla Pass, 5 Sept. 1965, *Findlay 240* (E). Near Pozanti, 31 July 1971, *Alice N2-844* (E). IÇEL: Taurus Mts., 1836, *Kotschy 472* (BM, W). Taurus Mts., near Güllek Tabiat, July 1853, *Kotschy 256A & 355C* (G, W). Taurus Mts., between Gulek-Boghas and Gulek-Maden, 23 Aug. 1855, *Balansa s.n.* (G). NIGDE: w. of Maden, 30 July 1969, *Darrah 338* (E). MUGLA: Sandras Dağ, near Köklüce, 23 July 1947, *Davis 13636* (E). Sandras Dağ, near Ağla, 25 July 1947, *Davis 13589* (E). DENIZLI: Honaz Dagh, 13 Aug. 1932, *Regel s.n.* (B, G). Baba Dagh, 18 Aug. 1932, *Regel s.n.* (B). Baba Dagh, above Kadiköy, 23 Aug. 1950, *Davis 18418* (E). Between Denizli and Tas Oçagi, 13 July 1947, *Davis 13230* (E). AYDIN: above Priene, 20 Aug. 1950, *Davis 18349* (E). IZMIR: Tmolus Mt., near Birgui, 23 July 1854, *Balansa 319* (E, G, W). MANISA: near Manisa, Dec. 1961, *Baytop s.n.* (Herb. Huber-Morath). BALIKESIR: Kazdağ, 9 July 1960, *Baytop 6031* (E). Marmara Adasi, 15 June 1968, *Baytop 13648* (E). ÇANAKKALE: Dardanelles, 1883, *Sintenis 456* (E). Bayramic Is., 16 Aug. 1951, *Akbas s.n.* (E). TEKIRDAĞ:

Ganos Dağ, 14 July 1968, *Baytop 13535* (E). ERDINE: between Erdine and Bosnaköy, 23 July 1968, *Baytop 14103* (E). SAKARYA: Sapanca, 8 Aug. 1962, *Davis 39172A* (E).

e. ssp. **virens** (Hoffmannsegg et Link) Ietswaart – **Figs. 29, 30** and **31.**

O. vulgare Linnaeus ssp. *virens* (Hoffmannsegg et Link) Ietswaart *stat. nov. O. virens* Hoffmannsegg et Link, Fl. Portugaise 1: 119 (1809); Willkomm & Lange, Prodr. Fl. Hisp. 2: 398 (1868); Boissier, Fl. Or. 4: 552 (1879); Lojacono Pojero, Fl. Sicula 2: 194 (1904); Jahandies & Maire, Cat. Pl. Maroc 3: 650 (1934); Tutin et al., Fl. Eur. 3: 171 (1972). *O. vulgare* Linnaeus var. *virens* (Hoffmannsegg et Link) Bentham, Lab. Gen. Sp.: 335 (1834). *O. virens* Hoffmannsegg et Link var. *genuinum* Coutinho, Fl. Port.: 612 (1913) – Type: *Moller s.n.*, Portugal, near Coimbra (neo. COI).

O. macrostachyum Hoffmannsegg et Link, Fl. Portugaise 1: 120 (1809). *O. vulgare* Linnaeus var. *macrostachyum* (Hoffmannsegg et Link) Brotero, Phyt. Lus. 2: 91 (1827). *O. virens* Hoffmannsegg et Link var. *macrostachyum* (Hoffmannsegg et Link) Coutinho, Fl. Port.: 612 (1913).

O. virens Hoffmannsegg et Link var. *spicatum* Rouy, Naturaliste 12: 93 (1882).

O. virescens Poiret, in Lamarck & Poiret, Encycl. Méth. Suppl. 4: 186 (1816).

O. silvestre Ortega ex Sampaio, Fl. Port.: 511 (1947).

Stems erect, up to 100 cm long, pilose to glabrescent (hairs c. 0.8 mm long). Branches up to 10 pairs per stem, 1.5 (0.2 – 10) cm long. *Leaves* up to 25 pairs per stem, (longly) ovate or oval, tops acute or $\pm$ obtuse, 17 (3 – 35) mm long, 9 (1.5 – 22) mm wide, pilose to glabrescent (hairs c. 0.5 mm long), sessile glands not conspicuous, up to 1000 per cm^2; margins entire or remotely serr(ul)ate; petioles up to 15 mm long. *Spikes* cylindrical or ovoid, 10 (5 – 35) mm long, 5 (4 – 8) mm wide. *Bracts* 7 (3 – 22) pairs per spike, (longly) (ob)ovate or oval, 6 (3.5 – 11) mm long, 4 (2 – 7) mm wide, membranous, usually glabrous, sometimes more or less pilosellous, usually yellowish green, sometimes slightly purple. *Calyces* 3.5 (3 – 5) mm long. *Corollas* 8.5 (7 – 11) mm long, white, seldom tinged pink. *Staminal filaments* up to 4.5 and 5.5 mm long. *Chromosome number* 2n = 30.

Geography and ecology. *O. vulgare* ssp. *virens* constitutes the western limit of the species. It occurs on the Azores, Canary Islands, Madeira, the Iberian Peninsula, the western part of North Africa and on the Balearic Islands. The occurrence on the first three (groups of) islands possibly is due to introduction by man. It is found on calcareous and non-limy hills and mountain slopes from 100 – 2200 m, sometimes on partly shaded places. It has been found flowering from May to August.

Notes. 1. On the islands mentioned as well as in southern France, southern Italy and in Greece intermediates occur between ssp. *virens* and ssp. *viride*. The geographical limit between the two sspp. has been drawn somewhat arbitrary. All specimens found in and north of the Pyrénées and east of the Balearic Islands are reckoned to be ssp. *viride*. In northern Spain intermediate forms occur between ssp. *virens* and ssp. *vulgare*. 2. Ssp. *virens* differs from the latter as well as from ssp. *viride* in the compact inflorescences, the usually larger, ($\pm$) glabrous, yellowish green bracts. 3. On several places hybrids have been found between ssp. *virens* and *O. majorana* (see p. 138).

Azores: Pico, 1838, *Hochstetter s.n.* (G, L, W).

Madeira: 1856, *Mason s.n.* (W). Jardin da Serra, 10 Aug. 1865, *Mandon s.n.* (G, W). Pico Grande, 1 Aug. 1900, *Bornmüller 1067* (JE).

Canary Is.: Palma, 1845, *Bourgeau 264* (G, W). Palma, 10 June 1896, *Kuegler s.n.* (JE). Tenerifa, La Laguna, 8 June 1900, *Bornmüller 1068* (G, W). Tenerifa, Orotava, el Montijo, Aug. 1921, *Burchard 38* (G).

Portugal: near Lisboa, July 1839, *Welwitsch 150* (G, L, W). Algarve, St. Pedro, 8 July 1853, *Bourgeau 1995* (G). Near Coimbra, June 1863, *Carvalho s.n.* (COI). Coimbra, June 1876, *Moller s.n.* (type). Coimbra, Aug. 1877, *Moller s.n.* (COI). Algarve, Alte, June 1878, *Moller s.n.* (COI). Near Lisboa, Sept. 1879, *da Cunha s.n.* (COI). Elvas, Arcos da Amorcira, Aug. 1879, *Silva Senna s.n.* (COI). Serra de Monsanto, Aug. 1879, *Daveau s.n.* (COI). Algarve, Faro, Aug. 1880, *Guimaraes s.n.* (COI). Near Lisboa, Aug. 1880, *da Silva s.n.* (COI). Cascaes, Aug. 1880, *Coutinho s.n.* (COI). Monte da Torre, June 1881, *da Cunha s.n.* (COI). Near Coimbra, July 1882, *Schultz 1444* (COI, JE, W). Leiria, July 1882, *Costa Lobo s.n.* (COI). Between Bemfica and Caneças, July 1883, *Daveau 651* (G, W). Bussaco, July 1883, *Loureiro s.n.* (COI). Coimbra, Cidral, June 1884, *Pereira da Silva s.n.* (COI). Cezimbra, June 1884, *da Silva s.n.* (COI). Pedro do Sul, July 1884, *Moller s.n.* (COI). Celarico da Beira, July 1885, *Ferreira s.n.* (COI). Arredoras da Guarda, Misarella, July 1885, *Ferreira s.n.* (COI). Serra da Estrella, July 1886, *Moller s.n.* (COI). Anredores de Torres Vedras, Aug. 1887, *de Barros e Cunha 659A* (COI). Coimbra, Balea, June 1888, *Moller 496* (COI). Near Vimioso, June 1888, *de Marir s.n.* (COI). Algarve, Loulé, July 1888, *Fernandes s.n.* (COI). Gouveia, July 1890, *Ferreira s.n.* (COI). Vermoil, July 1890, *Moller s.n.* (COI, G). Near Montejunto, Montegil, June 1892, *Moller s.n.* (COI). Serra do Gerez, Aug. 1892, *Moller s.n.* (COI). Povoa de Lanhoso, July 1894, *Sampaio s.n.* (COI). Natta do Fundad, July 1901, *Zimmerman s.n.* (COI). Near Castro Daire, Sept. 1901, *Heurigues s.n.* (COI). Louza, Flor da Rosa, Sept. 1909, *Santos s.n.* (COI). S. of Oporto, 7 Aug. 1910, *Johnston s.n.* (BM). Estremadura, Setubal, 4 June 1938, *Rothmaler 13471* (G, JE). Algarve, Monchique, 24 July 1938, *Zerny s.n.* (G). Tras-os-Montes, Vinhaes, 1 Aug. 1938, *Rothmaler 14027* (JE). Estremadura, Alcobaca, 5 Aug. 1938, *Rothmaler 14092* (JE). Monçao, rio Minho, 13 July 1945, *Garcia 705* (COI). Minho, Gerez, 15 July 1947, *Sinclair 4651* (BM). Minho, Monçao, 10 Sept. 1947, *da Silva 1103* (COI). Serra do Gerês, 9 July 1948, *Fernandes & Soeesa 2649* (COI). Soito do Bispo, near Guarda, 24 June 1950, *Fernandes & Matos 3484A* (COI). Coimbra, Camazao, 17 July 1957, *Matos & Marques s.n.* (COI). Pinhao on Douro, 29 July 1959, *Duvigneaud s.n.* (L). Near Nazaré, 7 June 1960, *Fernandes, Fernandes & Matos 7148* (COI). Miradonro da Serra da Boa Viagem, 3 June 1966, *Reis Moura 727* (COI).

Spain. Mainland: Granada, Guejar, 1837 *Boissier s.n.* (W). Sierra da Guadarrama, Aug. 1841, *Reuter s.n.* (BM, G). Sierra Nevada, Aug. 1848, *Funk s.n.* (W). Arragonia, between Bera and Borja, July 1850, *Willkomm 433* (W). Sierra Nevada, Guejar de la Sierra, 14 July 1851, *Bourgeau 1426* (G). Villafranca, Vierzo, 1851 – 52, *Lange s.n.* (JE). Granada, Lanjaron, July 1853, *Alioth s.n.* (G). Madrid, Pontón da la Oliva, 16 July 1858, *Isern s.n.* (M). Granada, Langeron, 12 July 1873, *Winkler 153* (JE). Near Granada, 26 July 1876, *Hackel s.n.* (W). Sierra de Mizas, 12 Aug. 1888, *Reverchon s.n.* (G). Ronda, 18 Aug. 1889, *Reverchon s.n.* (G, JE, L, W). Granada, Sierra Nevada, July 1891, *Porta & Rigo 576* (BM, W). Valencia, Sept. 1892, *Ségorbe* s.n. (G). Terual, Valacloche, Aug. 1893, *Reverchon s.n.* (G). Barrancon de Valentina, July 1904, *Reverchon s.n.* (G). Granada, Barranco del Orio Segura, July 1906, *Reverchon s.n.* (G). N. slopes of Sierra Nevada, 12 July 1927, *Bayer s.n.* (L). La Molata (Ciudad Real), 18 July 1934, *Albo s.n.* (M). Sierra de Corbera, June 1945, *Borja s.n.* (G). Mts. of Cazorla, 5 July 1948, *Heywood & Davis 183* (BM). Jaen, Sierra de Cazorla, Arrozo de los Cierzos, 22 July 1951, *Heywood 1523* (BM). Bocequillas s. of Burgos, 3 Aug. 1952, *de Wit 5215* (L). Toledo, between Talarera de la Reina and Arenas de San Pedro, 30 June 1972, *Gibbs & Dominguez D224* (E).

Balearic Is.: Foret de Castillo d'Arlaró, 15 June 1912, *Bianor 1427* (BM, G, JE).

Morocco: Moyen Atlas, Azrou, 8 Aug. 1924, *Jahandiez 911* (G).

f. ssp. **viride** (Boissier) Hayek – **Figs. 29, 30** and **31.**

O. vulgare Linnaeus ssp. *viride* (Boissier) Hayek, Prodr. Fl. Penins. Balc. 2: 334 (1931); Tutin et al., Fl. Eur. 3: 171 (1972). *O. vulgare* Linnaeus var. *viride* Boissier, Fl. Or. 4: 551 (1879); Fiori, Nuova Fl. Anal. Italia 2: 456 (1969). *O. viride* (Boissier) Halácsy, Consp. Fl. Graec. 2: 554 (1902); Rechinger, Fl. Aegaea: 531 (1943). – Type: *Bourgeau 700*, Turkey, Pontus Lazicus (holo. G).

O. heracleoticum Linnaeus, Sp. Pl.: 589 (1753). – Type: *Linnaeus 743.8* (holo. LINN).

O. minus Garsault, Traité Pl. Anim. 3: 257 (1767).
O. humile Miller, Gard. Dict. VIII Ed.: no. 4 (1768). *O. vulgare* Linnaeus var. *humile* (Miller) Bentham, Lab. Gen. Sp.: 335 (1834). – Type: *Miller s.n.* (holo. BM).
O. oblongatum Link, Enum. Pl. Horti Bot. Berol. 2: 114 (1822).
O. parviflorum Dumont d'Urville, Mem. Soc. Linn. Paris 1: 327 (1822).
O. normale Don, Prodr. Fl. Nepal.: 113 (1825). *O. vulgare* Linnaeus var. *normale* (Don) Briquet, in Engler & Prantl, Nat. Pflanzenfam. 4(3a): 309 (1895).
O. wallichianum Bentham, in Wallich, Num. List Dried Sp.: no. 1565 (1829). – Type: *Wallich 1565*, Nepal (holo. S).
O. vulgare Linnaeus var. *album* Fraas, Syn. Pl. Fl. Class.: 181 (1845).
O. vulgare Linnaeus var. *virens* Koch, Linnaea 19: 24 (1847).
O. virens Hoffmannsegg et Link var. *siculum* Bentham, in de Candolle, Prodr. 12: 193 (1848). *O. siculum* (Bentham) Nyman, Consp. Fl. Eur.: 592 (1881).
O. vulgare Linnaeus var. *smyrnaeum* Bentham, in de Candolle, Prodr. 12: 193 (1848).
O. albiflorum Koch, Linnaea 21: 662 (1848). *O. vulgare* Linnaeus var. *albiflorum* (Koch) Candargy, Bull. Soc. Bot. France 44: 459 (1897). – Type: *Koch 878*, Turkey, valley of Djimil (holo. G).
O. albiflorum Koch var. *congestum* Koch, Linnaea 21: 662 (1848).
O. angustifolium Koch, Linnaea 21: 661 (1848). – Type: *Koch s.n.*, cultivated (holo. B).
O. pruinosum Koch, Linnaea 21: 663 (1848).
O. viridulum Martrin-Donos, Fl. Tarn: 551 (1864). *O. vulgare* Linnaeus var. *viridulum* (Martrin-Donos) Briquet, Lab. Alpes Marit. 3: 483 (1895).
O. normale Don var. *incanum* Schmidt et Schlagintweit, J. Bot. 6: 234 (1868).
O. vulgare Linnaeus var. *magnilimbis* Boissier, Fl. Or. 4: 551 (1879). – Type: *Koch s.n.* (holo. G).
O. sardoum (Moris) Nyman, Consp. Fl. Eur.: 592 (1881).
O. vulgare Linnaeus var. *virescens* Cariot et St. Lager, Et. Fl.: 663 (1888).
O. semiglaucum Boissier et Reuter ex Briquet, Lab. Alpes Marit. 3: 484 (1895). – Type: *Boissier s.n.* France, near Lantosque (holo. G).
O. vulgare Linnaeus var. *semiglaucum* Boissier ex Briquet, Lab. Alpes Marit. 3: 484 (1895).
O. vulgare Linnaeus var. *laxiflorum* Post, Fl. Syr. Palest. Sin.: 617 (1896).
O. vulgare Linnaeus var. *longespicatum* Post, Fl. Syr. Palest. Sin.: 617 (1896).
O. gussonei Tineo ex Lojacono Pojero, Fl. Sicula 2: 195 (1904).
O. viride (Boissier) Halácsy var. *hyrcanum* Bornmüller, Beih. Bot. Centralbl. 33: 307 (1915). – Type: *Brüns 368*, Iran, Damavand Mts. (holo. B). *O. hyrcanum* Bornmüller, loc. cit., *nomen nudum*.
O. strobilaceum Mobayen et Gahraman, Bull. Soc. Bot. Fr. 125: 389 (1978).

Stems erect, sometimes ascending, up to 100 cm long, (densely) pilose(llous) to glabrescent (hairs c. 0.8 mm long), sometimes glaucous. Branches up to 20 pairs per stem, 2.5 (0.2 – 20) cm long. *Leaves* up to 35 pairs per stem, (longly) ovate or oval, sometimes roundish, tops acute to obtuse, 17 (2 – 55) mm long, 10 (2 – 30) mm wide, (densely) pilose(llous) (hairs c. 0.5 mm long), sometimes glaucous, sessile glands inconspicuous, up to 900 per cm^2; margins entire or remotely ser(rul)ate; petioles up to 15 mm long. *Spikes* ovoid or cylindrical, 6 (3 – 22) mm long, 4 (3 – 6) mm wide. *Bracts* 5 (2 – 17) pairs per spike, 5 (2 – 8) mm long, 2 (1 – 4) mm wide, (ob)ovate or oval, herbaceous or membranous, (densely) pilosellous to glabrescent, usually green, sometimes glaucous or slightly purple. *Calyces* 3 (2.5 – 4) mm long. *Corollas* 7 (5 – 9) mm long, white, seldom tinged pink. *Staminal filaments* up to 3.5 and 5 mm long.

Geography and ecology. *O. vulgare* ssp. *viride* has a large distribution area: it is found from Corse to East China. It grows from sea level up to 3000 m, in habitats that vary from dry to rather damp, from full sun to partly shaded. It occurs on various types of substrate, and flowers from May to October.

Notes. 1. Ssp. *viride* shades off into ssp. *gracile*, ssp. *virens* and ssp. *vulgare*. From the first it differs in the more robust habit, more hairy stems, leaves and usually bracts, and in the often less glandular punctate leaves. From the ssp. *virens* it is differing in the smaller spikes, and smaller, usually green, often hairy bracts. Ssp. *viride* differs from ssp. *vulgare* in the usually white (sometimes pinkish) flowers, in the usually smaller, green (sometimes tinged pink), often hairy bracts. 2. There are in fact no morphological differences between specimens with white flowers and green bracts incidentally occurring in populations of ssp. *vulgare* (in West and Central Europe) and specimens with the same characters forming (nearly) uniformely some of the populations of ssp. *viride* in Italy, Greece and Turkey.

FRANCE: near Montmelas, 16 Aug. 1876, *Gandoger 1055* (BM, L). Flavigny, 27 July 1912, *Desplantes s.n.* (L). CORSE: near Corte, June 1906, *Gysberger s.n.* (W).

ITALY. ISCHIA: 12 July 1832, *Splitberger s.n.* (L). CAPRI: Castiglione, Aug. 1859, *Haeckel s.n.* (JE). 15 July 1898, *Kuegler s.n.* (JE). SICILIA: Oct. 1841, *Parlatore s.n.* (G). 1860, *Citarda s.n.* (JE). Castelbuono, 17 July 1906, *Martelli s.n.* (G). Near Palermo, 1907, *Ross 663* (L). MAINLAND: Napoli, 1847, *Rabenhorst s.n.* (BM). Calabria, Reggio, near Gerace and St. Luca, June 1898, *Rigo 343 & 4291* (BM, G, JE).

YUGOSLAVIA: Dalmatia, 8 July 1927, *Korb s.n.* (W).

GREECE. MAINLAND: Thessalia, near Kalabaka, 21 July 1893, *Halácsy s.n.* (W). Near Thessaloniki, June 1901, *Adamovic s.n.* (W). Mt. Olympus, July 1927, *Handel-Mazzetti s.n.* (G, W). Ibid., 9 July 1928, *Dirowski s.n.* (W). Mt. Pindus, near Grevia, 16 Aug. 1934, *Guiol 2563* (BM). Near Grevia, 16 Aug. 1934, *Regel s.n.* (G). Mt. Timphi, near Vikos, 13 July 1958, *Rechinger 21142* (W). Makedhonia, near Pangaeon, 22 June 1959, *Stainton 7747* (W).

TURKEY: Pontus Lazicus, 25 July 1862, *Bourgeau 700* (type). Topouzlubend, 22 June 1890, *Aznavour s.n.* (G). Paphlagonia, Kastambuli, 9 July 1892, *Sintenis s.n.* (JE). Nerzifoun, 1904, *Manissadjian s.n.* (G). KONYA: Ak Çhehir, 8 July 1907, *Saint-Lager s.n.* (G). ISPARTA: Sutçuler, Kuyaçuk Dağ, 30 July 1949, *Davis 15893* (E, W). Sutçuler, Aug. 1949, Dedegöl Dağ, *Davis 15930* (E). Dedegöldağ, 17 July 1968, *Sorger 68 – 43 – 49* (Herb. Sorger). BURSA: near Bursa, Oct. 1867, *Ball s.n.* (E). E. of Bursa, 19 July 1968, *Sorger 68 – 50 – 9* (Herb. Sorger). ISTANBUL: near Istanbul, 1 July 1844, *Noei 209* (W) Therapsia, 1889, *Aznavour s.n.* (G). Kandilli on Bosporus, 1909, *Wimmer 155* (W). Above Çubuklu, 16 July 1939, *Post s.n.* (G). SAKARYA: Dogançay, 1 July 1962, *Davis 36327* (E). BOLU: slope on Abant Gölü, 25 Sept. 1957, *Wagenitz & Beug 243* (B). ZONGULDAK: between Zonguldak and Çaycuma, 17 July 1962, *Davis 37633* (E). Between Yenice and Balikisik, 20 July 1962, *Davis 37911* (E). KASTAMONU: Ilgaz Dağlari, 11 Aug. 1960, *Khan et al. 683* (E). N. side of Ilgaz Dağ, 28 July 1962, *Davis 38430* (E). SINOP: between Gerze and Boyabat, 7 Sept. 1954, *Davis 25013* (BM, E). Between Sinop and Gerze, 2 July 1958, *Huber-Morath 15553* (Herb. Huber-Morath). AMASYA: Sanadagh, Akdagh etc., 11 July 1889, *Bornmüller 1432* (W). Akdagh, 10 May 1896, *Foerster s.n.* (B, E, G). Between Erbaa and Kocak, 5 July 1967, *Tobey 2178* (E). SAMSUN: Lâdik, 26 June 1965, *Tobey 1179* (E). ORDU: between Fatsa and Aybast, 21 July 1965, *Tobey 1328* (E). W. of Ünye, 6 July 1969, *Sorger 69 – 21 – 1A* (Herb. Sorger). GIRESUN: near Yavuzkemal, 13 Aug. 1952, *Davis 20773* (BM, E). Between Kulakkaya and Giresun, 2 July 1955, *Huber-Morath 14167* (Herb. Huber-Morath). S. of Giresun, 7 July 1969, *Sorger 69 – 23 – 31* (Herb. Sorger). GÜMÜŞANE: Szanschak Gümüschkhane, above Artabir, 31 July 1894, *Haussknecht 7003* (E). Gumush Hane, 17 July 1934, *Balls 1721* (E). N. of Torul, 31 July 1969, *Buttler 14332* (Herb. Buttler). TRAPZON: 6 July 1906, *Handel-Mazzetti 198* (G). Near Trapzon, 1931, *Görz S12* (Herb. Huber-Morath). Between Sürmene and Of, 11 July 1958, *Huber-Morath 15554* (Herb. Huber-Morath). RIZE: Lazistan, valley of Djimil, Aug. 1866, *Balansa 1523* (E, W). Between Karadere and Guneyçe, 23 Aug. 1952, *Davis 20839* (E). Between Rize and Çayeli, 6 Aug. 1957, *Davis 32041* (BM, E). ÇORUH: above Artvin, 25 June 1957, *Davis 30037* (BM, E). Above Hopa, 16 Aug. 1957, *Davis 32413* (BM, E).

IRAN: Rasht, 5 June 1915, *Pravitz 947* (W). Khorasan, between Budjnurd and Morawe Tappeh, July 1937, *Rechinger 1874* (W). Mazanderan, valley of Talar, between Abbasabad and Cahi, 4 Aug. 1937, *Rechinger 2018* (W). Mazanderan, between Babalsar and Noshar, 5 Aug. 1937, *Rechinger 2069* (W). Gurgan, 12 July 1940, *Koelz 16155* (W). Gorgan, 30 Sept. 1940, *Koelz 16976* (W). Mazanderan, in valley of Talar, between Po-e Sefid and Sorkhabad, 20 June 1948, *Rechinger 5564* (G, W). Mazanderan, Kudjur, between Sanus and Kindj, Aug. 1948, *Rechinger 6586* (G, W). Gorgan, Saffiabad near Behshahr, 13 Sept. 1948, *Aellen 799* (G, W). Mazanderan, valley of Haraz, above Emarat, 26 July 1959, *Wendelbo 1499* (E, W). E. Azerbaijan, w. side of Hasi Amir Pass, 21 July 1964, *Grant 16309* (W). Ibid., 21 July 1964, *Grant 16379* (W). Mazandaran, Chalus river gorge, s. of Chalus, 26 July 1964, *Grant 16489* (W). Mazandaran, Caspian sea, w. of Mahmudabad, 26 July 1964, *Grant 16499* (W). Azerbaijan, pass between Astara and Ardabil, 27 July 1967, *Walton 64* (E). Khorasan, near Tangara, 21 Aug. 1967, *Walton 141* (E). Mazandaran, between Kalardasht and Vendarben, 16 Sept. 1970, *Termé 14472E* (W). Gorgan, e. of Tangar, 18 July 1971, *Edmondson 715* (E). Gorgan, Golestan valley, 30 July 1974, *Wendelbo & Cobham 14328* (W). Mazandaran, Caspian sea, 30 Aug. 1974, *Wendelbo & Assadi 14585* (W).

AFGHANISTAN: Konar, Dewagal Darrah, valley w. of Chalas, pass to Darrahe Mazar, 31 Aug. 1973, *Anders 11096* (W).

PAKISTAN: Swat, 18 Aug. 1955, *Rahman 152* (W). Swat, Ushu, 22 Aug. 1962, *Rechinger 19465* (W). Malakand, Malakand, 1 June 1965, *Rechinger 30454* (W).

INDIA: Kamoon, Sept. 1836, *Wallich 1564* (W). Between Tsámba and Padri Pass, July 1856, *Schlagintweit 3589* (L). N. of Simla, Sept. 1864, *Stohtzka s.n.* (W). 1965, *Falconer s.n.* (L, W).

CHINA: Shao lin ssu, near Teng fong, Honan, Aug. 1907, *Schindler 160* (W). Tchouan, June 1912, *Maire s.n.* (G). Yünnan, Dali, 29 July 1914, *Meller s.n.* (W). Chansi, Weitze ping, 14 Aug. 1916, *Licent 1502* (W). Chi-Gung-Shan, n. of Hankow, July 1923, *Uhlenbroek s.n.* (L). Huangshan Anhuei, July 1926, *Chien 1294* (L). Anhwei, June 1936, *Fan & Li 173* (L).

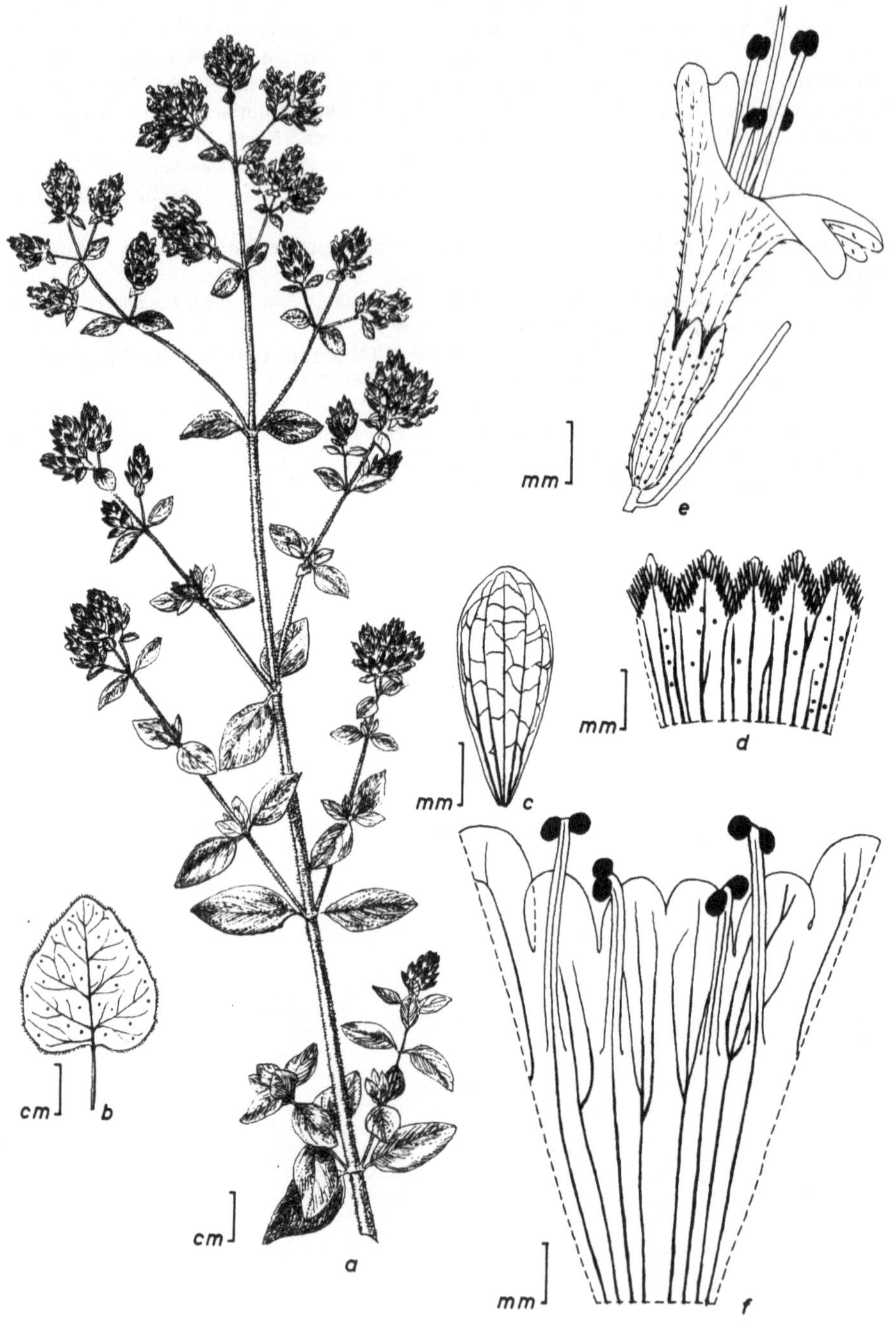

Figure 28. O. vulgare ssp. *vulgare:* a. habit; b. leaf; c. bract; d. calyx cut through the lower lip; e. flower with bract in side view; f. corolla cut through the lower lip.

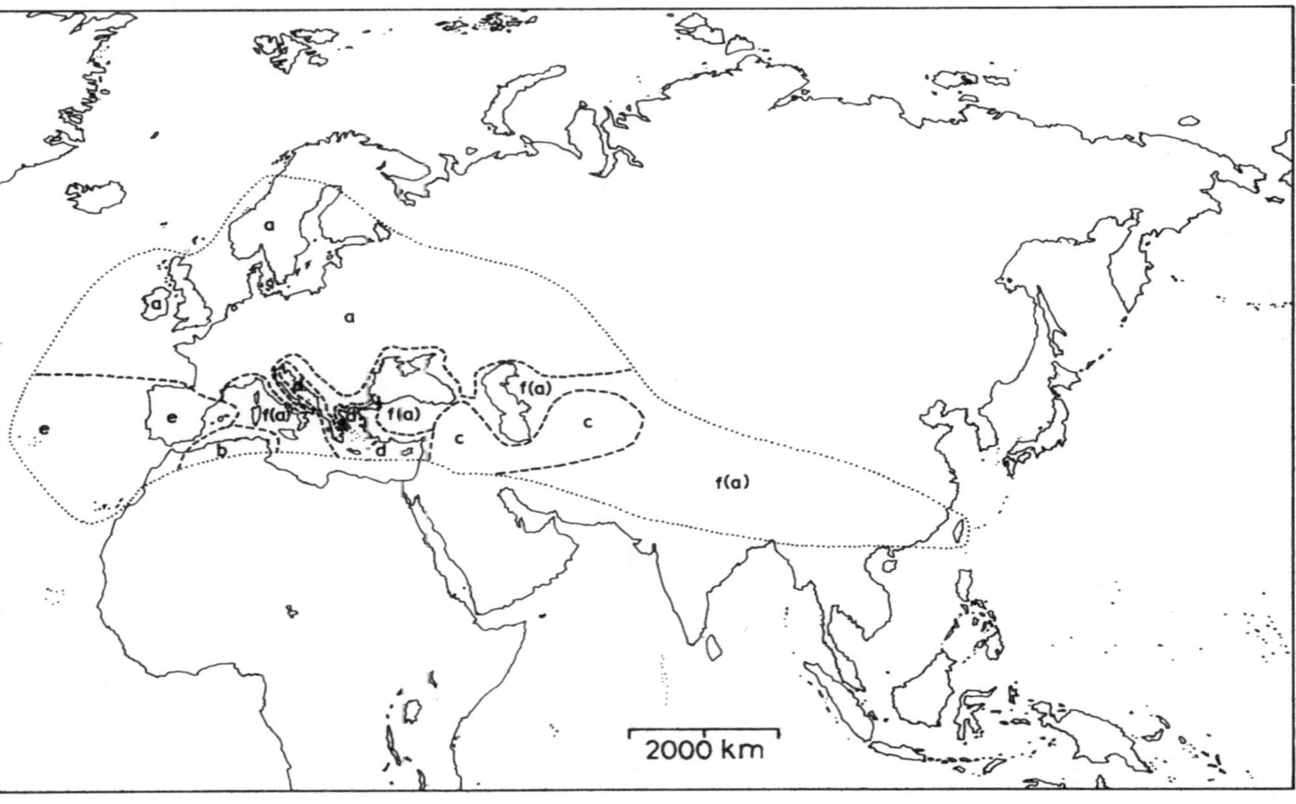

Figure 29. Distribution of *O. vulgare* .... and its subspecies: a. ssp. *vulgare*; b. ssp. *glandulosum*; c. ssp. *gracile*; d. ssp. *hirtum*; e. ssp. *virens*; f(a) mainly ssp. *viride*, occasionally ssp. *vulgare*.

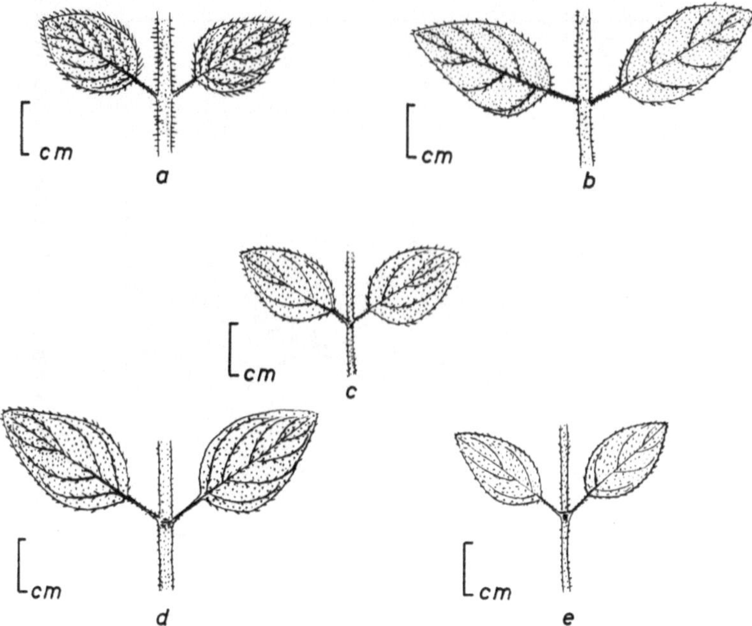

Figure 30. Stems with leaves of the subspecies of *O. vulgare*, except ssp. *vulgare* (for which see figure 28): a. ssp. *glandulosum;* b. ssp. *gracile;* c. ssp. *hirtum;* d. ssp. *virens;* e. ssp. *viride*.

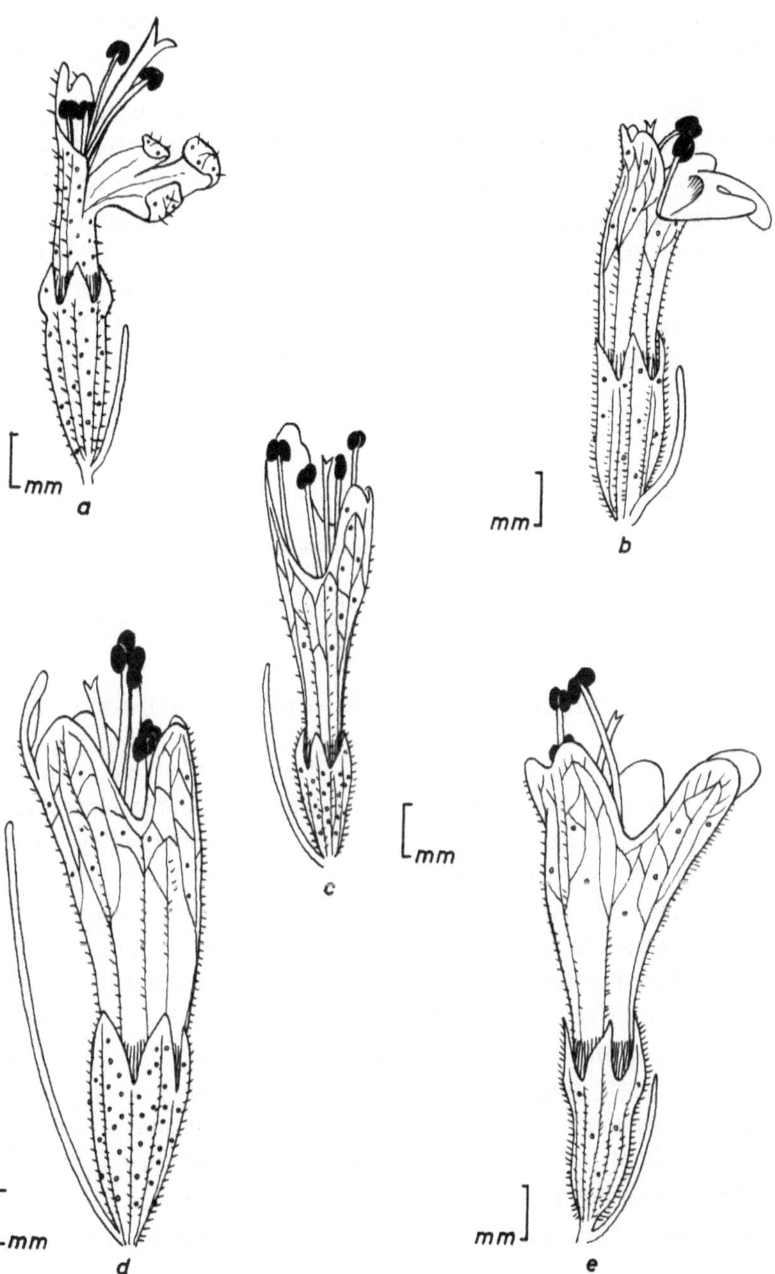

Figure 31. Flowers with bracts in side view of the subspecies of *O. vulgare*, except ssp. *vulgare* (for which see figure 28): a. ssp. *glandulosum;* b. ssp. *gracile;* c. ssp. *hirtum;* d. ssp. *virens*, e. ssp. *viride*.

X. Section Prolaticorolla Ietswaart

Section *Prolaticorolla* Ietswaart, Not. R. B. G. Edinburgh 38(1): 47 (1980). – Type: *Origanum laevigatum* Boissier.

Branches of the first order always present, those of the second order usually so, and those of the third order sometimes so. *Leaves* herbaceous or somewhat leathery. *Spikes* dense or somewhat loose, distinct, usually medium sized, erect. *Bracts* different from the leaves in shape, size, texture and colour, membranous, purple or greenish, (densely) imbricate, c. $\frac{3}{4}$ – $1\frac{1}{4}$ x calyces. *Flowers* 2 per verticillaster, bisexual or female, large. *Calyces* tubular with 5 (nearly) equal teeth for c. $\frac{1}{3}$, when fruiting teeth more or less contracted; throats pilose. *Corollas* 2-lipped for c. $\frac{1}{6}$, c. 3 x calyces. *Stamens* unequal in length, straight, (sub)included or shortly protruding; filaments c. $\frac{1}{4}$ x corollas.

35. Origanum compactum Bentham – Figs. 32 and 33.

O. compactum Bentham, Lab. Gen. Sp.: 334 (1834); Willkomm & Lange, Prodr. Fl. Hisp. 2: 398 (1868); Boissier, Fl. Or. 4: 551 (1879); Holmes, Perf. Ess. Oil Rec. 4: 73 (1913); Jahandiez & Maire, Cat. Pl. Maroc 3: 651 (1934); Coutinho, Fl. Port.: 611 (1939); Nègre, Petite Fl. Maroc Occ. 2: 182 (1962); Tutin et al., Fl. Eur. 3: 171 (1972). – Type: *Salzmann s.n.*, Morocco, near Tanger, 1831 (holo. K).

O. glandulosum Salzmann ex Bentham, Lab. Gen. Sp.: 335 (1834). – Type: *Salzmann s.n.*, Morocco, near Tanger, 1831 (holo. K).

O. creticum Schousbou ex Ball, Journ. Linn. Soc. 16: 610 (1878). – Type: *Schousbou 107* (holo. G, iso. W).

Woody perennials. Stems erect or ascending, up to 75 cm long, light or dark brown, hirsute (hairs c. 2 mm long). Branches of the first order present, in the upper $\frac{2}{5}$ – $\frac{4}{5}$ of the stems, up to 18 pairs per stem, 1.0 (0.2 – 12) cm long; branches of the second order usually present and those of the third order seldom so, both very short. *Leaves* up to 23 pairs per stem, shortly petiolate (petioles up to 15 mm long), ovate or oval, margins remotely serrulate or entire, tops usually $\pm$ acute, 20 (4 – 30) mm long, 10 (2 – 20) mm wide, herbaceous, light green, pilose (hairs c. 1 mm long), sessile glands up to 1800 per cm^2. *Spikes* cylindrical or ovoid, 15 (5 – 50) mm long, 6 (5 – 7) mm wide. *Bracts* 8 (4 – 30) pairs per spike, obovate or oval, tops acuminate, 8 (6 – 11) mm long, 3.5 (2 – 5) mm wide, partly purple, pilosellous. *Flowers* (sub)sessile. *Calyces* 4 (3 – 4.5) mm long, outside pilosellous; teeth 1.2 (0.8 – 1.4) mm long. *Corollas* 11 (9 – 14) mm long, pink (or white), outside pilosellous; upper lips divided, for c. $\frac{1}{5}$, into 2, c. 0.5 mm long lobes; lower lips divided, for c. $\frac{1}{2}$, into 3, slightly unequal, 1.5 (1 – 2) mm long lobes. *Staminal filaments* up to 4.5 and 6 mm long. *Styles* up to 16 mm long.

Geography and ecology. *O. compactum* is found in the southern most part of Spain and the adjacent regions of Morocco. It grows on dry hills, up to c. 700 m, sometimes between trees and shrubs. It has been found flowering from June to August.

Notes. 1. *O. compactum* differs from *O. ehrenbergii* in its larger, purple bracts, and from *O. laevigatum* in its villous stems and larger bracts. From *O. vulgare* ssp. *virens*, *O. compactum* differs in its larger corollas with relatively shorter lips and stamens, and also in its larger, pilose, purple bracts and usually pink flowers. 2. For a possible hybrid with *O. compactum* as one parent see notes for *O. grosii* (p. 101).

MOROCCO: near Tanger, 1831, *Salzmann s.n.* (type). Near Tanger, June 1831, *Schousbou 107* (G, W). Ahouana near Zuiet, 21 April 1912, *Pitard 1625* (G). Chaouia, Boulhaut, 19 June 1912, *Pitard 1626* (G). Oued Djedidah near Neknès, 1 July 1923, *Jahandiez 633* (G). Near Xauen, c. 600 m, 23 June 1928, *Font Quer 351* (G). Chaouia, El Aïoun, near Boulhaut, 18 June 1937, *Gattefossé 983* (G, JE).
SPAIN: Andalusia, Alcala de los Gazules, 2 Aug. 1853, *Bourgeau s.n.* (G).

36. **Origanum ehrenbergii** Boissier – **Figs. 32** and **33.**

O. ehrenbergii Boissier, Fl. Or. 4: 551 (1879); Post & Dinsmore, Fl. Syr. Palest. Sin. 2: 333 (1933); Thiebaut, Fl. Libano-Syr. 3: 46 (1953). – Type: *Ehrenberg s.n.*, Lebanon, Mt. Sanin (holo. G).
O. ehrenbergii Boissier var. *parviflorum* Bornmüller, Beih. Bot. Centralbl. 31: 248 (1914). – Type: *Bornmüller 12323*, Lebanon, Lebanon Mts., near Ain Zahalta (holo. B).

Woody perennials. Stems up to 85 cm long, usually erect, light brown, hirsute at the bases (hairs c. 2 mm long, to the top of the stems deminishing in length and density). Branches of the first order always present, in the upper $\frac{1}{2}$ of the stems, up to 10 pairs per stem, 1 (0.3 – 7) cm long; those of the second order usually present and those of the third order sometimes so. *Leaves* up to 25 pairs per stem, shortly petiolate (petioles up to 7 mm long), ovate or oval, tops ± acute, 20 (4 – 30) mm long, 10 (2 – 18) mm wide, herbaceous, light green, sometimes more or less glaucous, pilose to ± glabrous (hairs c. 0.8 mm long), sessile glands up to 2500 per cm^2. *Spikes* often somewhat loose, cylindrical, 15 (5 – 35) mm long, c. 5 mm wide. *Bracts* 8 (3 – 15) pairs per spike, obovate or oval, tops acuminate or acute, 5 (4 – 5) mm long, 2 (1.5 – 2.5) mm wide, (nearly) glabrous, greenish. *Flowers* (sub)sessile. *Calyces* 4.8 (4 – 5) mm long, outside pilosellous; teeth 1.3 (0.9 – 1.5) mm long. *Corollas* 11 (7 – 14) mm long, white, outside pilosellous; upper lips divided, for c. $\frac{2}{5}$, into 2, c. 1 mm long lobes; lower lips divided, for c. $\frac{3}{5}$, into 3, slightly unequal, 1.4 (1 – 2) mm long lobes. *Staminal filaments* up to 3.5 and 4.5 mm long. *Styles* up to 16 mm long.

Geography and ecology. *O. ehrenbergii* occurs in a few places in the Lebanon Mts., where it grows up to c. 1500 m, sometimes under pine trees. It has been found flowering in June and July.
Notes. 1. In the type specimen of *O. ehrenbergii* var. *parviflorum* small flowers with reduced stamens are found, so most probably here a female specimen is involved. 2. *O. ehrenbergii* differs from the related species *O. laevigatum* in its villous stems, its greenish bracts and white corollas. It is easily distinguishable from *O. compactum* by its smaller, greenish bracts. 3. *O. ehrenbergii* hybridizes in nature with *O. syriacum* (see p. 135).

LEBANON: Lebanon Mts., between Bekfaia and Zahlé, on sandy places, 5 July 1853, *Gaillardot 521* (G, JE). Merdj Besri near Saïda, 22 June 1854, *Gaillardot s.n.* (JE). Lebanon Mts., environs of Brummana, July 1879, *Cramer s.n.* (G). Beitneri, 27 July 1879, *Peyron 689* (G). Beitneri, 25 July 1881, *Peyron s.n.* (G). Lebanon Mts., Nellanssa, 1895, *Tanus Botos s.n.* (JE). Lebanon Mts., near Brummana, in the lower zones, under pine trees, c. 750 m, June 1897, *Bornmüller 1249* (G, JE, W). Lebanon Mts., near Ain Zahalta, on w. slopes, under pine trees, 20 June 1910, *Bornmüller 12323* (B). Near Rekka, 15 July 1920, *Cotto s.n.* (L).

37. **Origanum laevigatum** Boissier – **Figs. 33** and **34.**

O. laevigatum Boissier, Ann. Sc. Nat. 4(2): 252 (1854); Boissier, Fl. Or. 4: 550 (1879); Post & Dinsmore, Fl. Syr. Palest. Sin. 2: 333 (1933); Thiebaut, Fl. Libano-Syr. 3: 46 (1953). – Type: *Tchihatcheff s.n.*, Turkey, Cataonia and Antitaurus (holo. G).

O. laevigatum Boissier var. *laxum* Post, Fl. Syr. Palest. Sin.: 616 (1896). – Type: *Post s.n.*, Lebanon, Aintab (holo. BM).

Subshrubs. Stems up to 70 cm long, usually ascending and rooting at the bases, (dark or purplish) brown, nearly glabrous (at most a few, c. 0.2 mm long, hairs present, mainly at the bases), glaucous. Branches of the first order always present, in the upper $\frac{1}{5}$ – $\frac{3}{5}$ of the stems, up to 12 pairs per stem, 3.5 (0.5 – 17) cm long; those of the second order often present, those of the third order sometimes so. *Leaves* up to 20 pairs per stem, shortly petiolate (petioles up to 8 mm long), ovate or oval, tops obtuse or acute, 17 (3 – 30) mm long, 8 (1.5 – 17) mm wide, somewhat leathery, glaucous, often purplish, glabrous, sometimes scabrous (hairs c. 0.2 mm long), sessile glands hardly visible and nearly only present at the under sides, up to 1200 per cm^2. *Spikes* often loose, more or less ovoid or cylindrical in outline, 8 (5 – 20) mm long, 4 (3 – 6) mm wide. *Bracts* 4 (2 – 8) pairs per spike, ± lanceolate, tops acute, 4.5 (3 – 6) mm long, 1.2 (0.5 – 2) mm wide, glabrous, (more or less) purple. *Flowers* (sub)sessile. *Calyces* 5 (3 – 6) mm long, outside glabrous (sometimes teeth slightly pilosellous); teeth 1.5 (1 – 2) mm long. *Corollas* 12 (8 – 16) mm long, purple, outside pilosellous; upper lips divided, for c. $\frac{1}{2}$, into 2, 0.8 (0.5 – 1.2) mm long lobes; lower lips divided, for c. $\frac{3}{5}$, into 3, slightly unequal, 1.5 (0.5 – 2) mm long lobes. *Staminal filaments* up to 3 and 4.5 mm long. *Styles* up to 17 mm long.

Geography and ecology. *O. laevigatum* is known from several places in the Amanus Mts. (southern Turkey) and the adjacent mountains north and south of this range. It also has been found on Cyprus. It grows from 300 – 2000 m, sometimes in maquis and open woods. It flowers from April to September.
Notes. 1. The type of corolla found in the section *Prolaticorolla* is rather similar to that of *O. leptocladum* and some other species in the (not closely related) section *Brevifilamentum*. 2. *O. laevigatum* differs from *O. compactum* and *O. ehrenbergii* in its (nearly) glabrous stems and leaves. 3. In nature two hybrids have been found, of which *O. laevigatum* is one of the parental species. The other parents are *O. amanum* and *O. syriacum* (see pp. 135 and 140).

TURKEY. PROV. HATAY: in southern Cataonia and in Antitaurus, *Tchihatcheff s.n.* (type). Between Alexandretta and Belen, 1865, *Haussknecht s.n.* (JE). Alexandretta, on mountains above this place, Dec. 1865, *Haussknecht s.n.* (JE). Aintab, June 1882, *Post s.n.* (BM). Amanus Mts., Aug. 1891, *Post s.n.* (BM). Amanus Mts., c. 750 m, Aug. 1906, *Haradjian 491* (G, W). Amanus Mts., region of Hasanbeyli, c. 900 m, June 1908, *Haradjian 2621* (G, W). Amanus Mts., in valleys, 250 – 900 m, Sept. 1908, *Haradjian 4645* (G, W). Amanus Mts., Djebel Momsa near Antiocha, 600 – 1200 m, July 1909, *Haradjian 3200* (G, W). PROV. SEYHAN: Dildil Dag, 1500 – 2000 m, April 1911, *Haradjian 3904* (G, W). Dildil Dag above Haruniye, in maquis and open woodland, c. 450 m, 26 Aug. 1949, *Davis 16371* (G).

ADDENDUM

After the completion of the manuscript a new *Origanum* species from Cyrenaica was described, which is appended here. The species belongs to the Section *Anatolicon*.

38. Origanum pampaninii (Brullo et Furnari) Ietswaart.

O. pampaninii (Brullo et Furnari) Ietswaart *comb. nov. Amaracus pampaninii* Brullo et Furnari, Webbia 34(1): 439 (1979). – Type: *Brullo et Furnari s.n.*, Libya, Cyrenaica, Wadi Gattara near Barce (holo. CAT).

Subshrubs, flowers bisexual. Stems sparsely branched at the bases, erect or ascending, up to 40 cm long, light brown, more or less tomentose (hairs c. 1 mm long). Branches of the first order present, in the upper $\frac{2}{5}$ (seldom $\frac{1}{2}$), c. 8 pairs per stem, c. 0.3 cm long; sometimes a few, utmost short branches of the second order present. *Leaves* c. 17 pairs per stem, (sub)sessile (in the lower nodes with up to 1 mm long petioles), roundish or ovate, tops obtuse, 10 (9 – 16) mm long, 7 (2 – 12) mm wide, thin, greyish, more or less tomentose (hairs c. 1 mm long), sessile glands not obvious, up to 450 per cm^2. *Spikes* subglobose or ellipsoid, 9 (6 – 16) mm long, 6 (4 – 11) mm wide, more or less nodding. *Bracts* c. 5 pairs per spike, oblong or oval, tops more or less acute, 5 (5 – 6) mm long, c. 3 mm wide, green or partly slightly purplish, indumentum ± as in the leaves. *Calyces* 2-lipped for c. $\frac{1}{5}$, c. 5 mm long, outside and throats pilose; upper lips denticulate or (sub)entire; lowers lips approximately as long as the upper lips, consisting of 2 triangular, c. 1 mm long teeth. *Corollas* 2-lipped for c. $\frac{1}{5}$, c. 10 mm long, white or slightly pink, not saccate, outside somewhat pilosellus; upper lips divided, for c. $\frac{1}{4}$, c. 0.4 mm long lobes; lower lips divided for c. $\frac{3}{5}$, into 3, somewhat unequal, c. 1.5 mm long lobes. *Stamens*, the upper 2 included, the lower 2 shortly protruding; filaments up to 2 and 4 mm long. *Styles* often curved, up to 14 mm long.

Geography and ecology. *O. pampaninii* occurs in Cyrenaica somewhat east from the distribution area of the other 2 endemic *Origanum* species. It grows on the rocks faces of the Gebel near Barce which consists of limestone (just as the sites of the other 2 species). For the moment this is the only site where it is known from. It has been found flowering in May and June.

Notes. 1. Recently two new *Origanum* species (*O. akhdarense* and *O. pampaninii*) have been found in Cyrenaica, while the species known from there since 1913 (*O. cyrenaicum*) could not be recovered. It is questionable whether hybridization played a part in this. The situation is reminiscent of that found for *O. micranthum* (see *O. micranthum* x *vulgare* ssp. *hirtum*). 2. *O. pampaninii* differs from *O. cyrenaicum* amongst others in its calyces which are 2-lipped for only $\frac{1}{5}$, and which possess long lower lips. From *O. akhdarense* (to which it is closely akin) it differs in the following characters: more or less tomentose stems and leaves, (sub)sessile leaves, somewhat larger flowers, white or slightly pink corollas which are 2-lipped to c. $\frac{1}{5}$, short stamens of which the upper 2 are included in and the lower 2 shortly protruding from the corolla. 3. In the corolla characters just mentioned *O. pampaninii* is alike species in the section *Brevifilamentum*. 4. According to Brullo & Furnari *O. pampaninii* flowers exclusively in spring, and emits a strong smell.

LIBYA. CYRENAICA: Wadi Gattara near Barce, 21 May 1974, *Brullo & Furnari s.n.* (type).

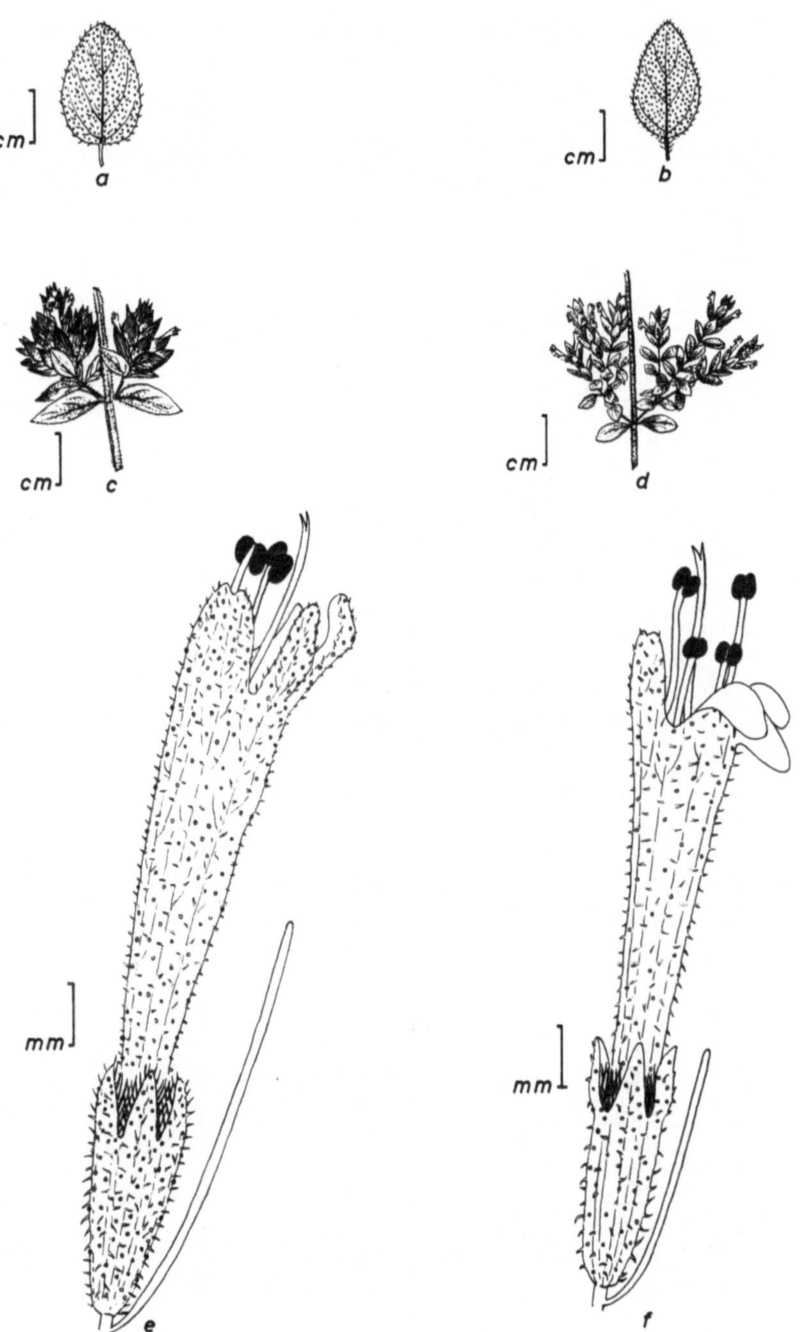

Figure 32. Leaves, spikes, and flowers with bracts in side view of the species in the section *Prolaticorolla*, except *O. laevigatum* (for which see figure 34): a., c. and e. *O. compactum*; b., d. and f. *O. ehrenbergii*.

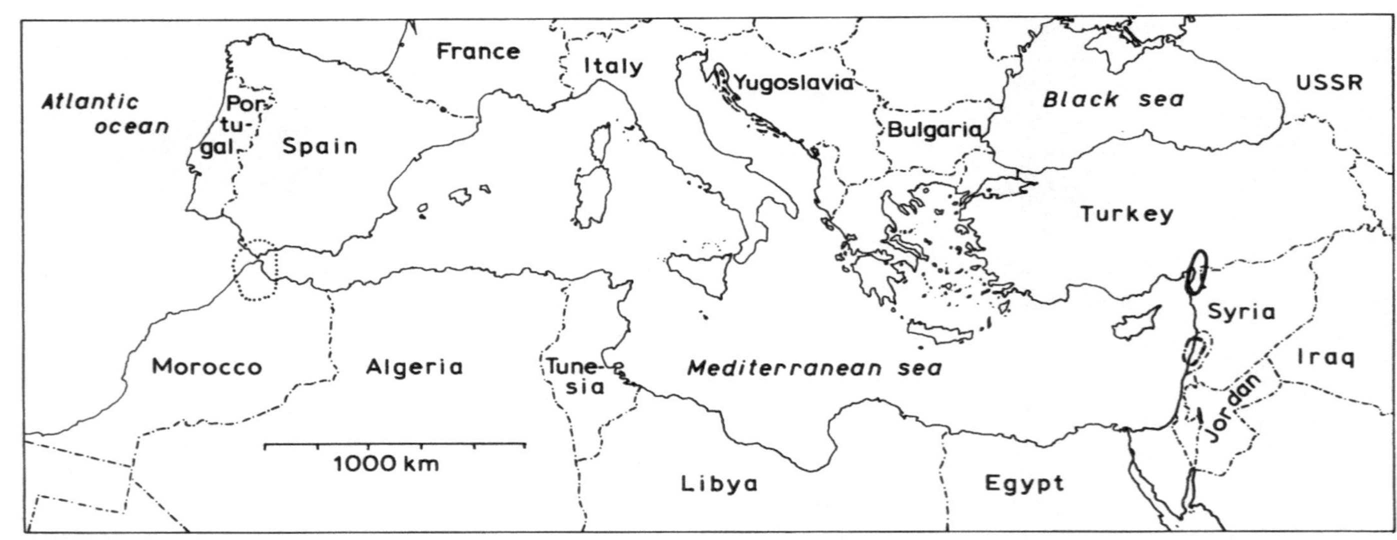

Figure 33. Distribution of the species in the section *Prolaticorolla*: .... *O. compactum*; _._._. *O. ehrenbergii*; ______ *O. laevigatum*.

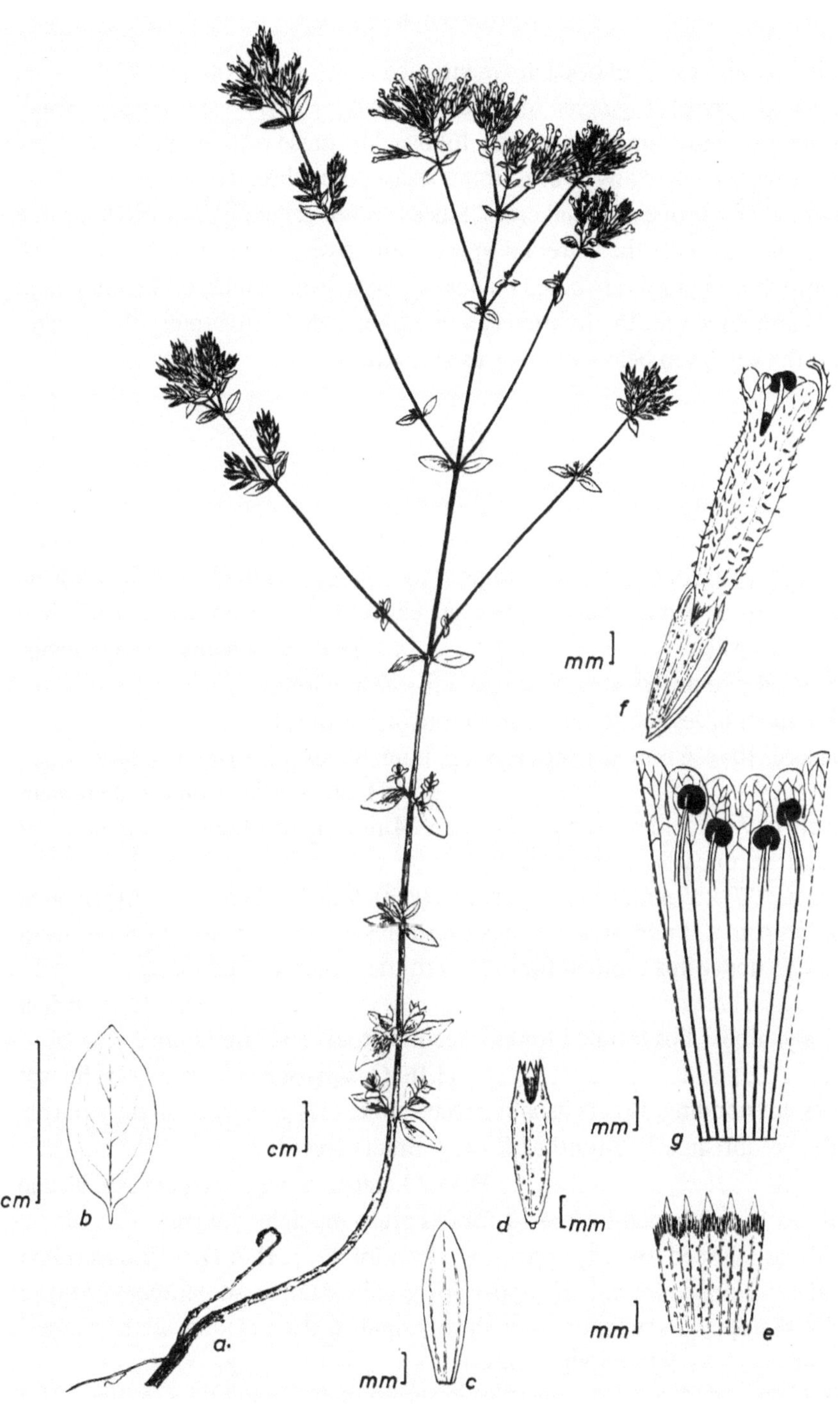

Figure 34. O. laevigatum: a. habit; b. leaf; c. bract; d. calyx; e. calyx cut through the lower lip; f. flower with bract in side view; g. corolla cut through the lower lip.

II.5. HYBRIDS

Introduction

In this chapter all known hybrids are mentioned which have both parental species belonging to the genus *Origanum*. In one case possibly a *Thymus* species is involved. First the hybrids with names are treated, followed by those without names. No new names have been given to any hybrid from the last group, because they are putative, or only known from one artificial cross. Since it is not generally known which one was the pollen parent, the parental species are given in alphabetic order. All descriptions of the hybrids are based on one or only a few specimens. This combined with the knowledge that the hybrids can be rather variable (chapter I.11), makes clear that the key given below must be used carefully.

Key

1. Calyces c. 6 mm long. Corollas 11 – 20 mm long
  2. Bracts c. 7 × 2 mm. Calyx upper lips toothed for c. $\frac{4}{5}$; teeth c. 2 mm long; lower lips ± as long as the upper lips .... **H.4. O. × dolichosiphon**
  2. Bracts c. 10 × 8 mm. Calyx upper lips entire; lower lips much shorter than the upper lips .... **H.13. O. amanum × dictamnus**
1. Calyces c. 4 mm or less long. Corollas 2.5 – 12 mm long
  3. Spikes more or less nodding. Bracts more or less purple
    4. Bracts c. 10 × 6 mm. Calyces c. 6 mm long. Stems more or less lanate .... **H.14. O. calcaratum × dictamnus**
    4. Bracts c. 7 × 5 mm or less. Calyces c. 4 mm or less long. Stems more or less pilose, pilosellous or hirtellous
      5. Bracts c. 7 × 5 mm. Calyx upper lips (sub)entire ... **H.5. O. × hybridinum**
      5. Bracts c. 6 × 3 mm or less. Calyx upper lips with triangular or deltoid teeth
        6. Calyx upper lips toothed for c. $\frac{4}{5}$; teeth triangular, c. 1 mm long .... **H.8. O. × lirium**
        6. Calyx upper lips toothed for c. $\frac{1}{6}$; teeth deltoid, c. 0.2 mm long .... **H.16. O. sipyleum × vulgare** ssp. **hirtum**
  3. Spikes not nodding. Bracts usually green, sometimes greyish or slightly purple
    7. Calyces subregularly 5-toothed for c. $\frac{1}{3}$. Bracts greyish .... **H.15. O. micranthum × vulgare** ssp. **hirtum**
    7. Calyces 1- or 2-lipped for c. $\frac{1}{5}$ – $\frac{4}{5}$. Bracts green or slightly purple
      8. Calyces 1-lipped for c. $\frac{4}{5}$; upper lips subentire ..... **H.7. O. × intermedium**
      8. Calyces 2-lipped for c. $\frac{1}{5}$ – $\frac{2}{5}$; upper lips usually dentate, sometimes subentire
        9. Calyx upper lips subentire or with very small, c. 0.2 mm long, teeth or lobes. Corollas 5.5 – 9 mm long, pinkish
          10. Calyces c. 4 mm long; teeth of lower lips c. 0.7 mm long. Corollas 7 – 9 mm long .... **H.11. O. × pabotii**

10. Calyces c. 2.2 mm long; teeth in lower lips c. 0.2 mm long. Corollas c. 5.5 mm long .... **H.10. O. × minoanum**

9. Calyx upper lips toothed for c. $\frac{1}{5}$ – $\frac{1}{2}$; teeth 0.3 – 0.6 mm long. Corollas 2.5 – 12 mm long, pink(ish) or white

11. Corollas 9 – 12 mm long. Calyx lower lip teeth somewhat shorter to ± as long as the upper lips, c. 1 mm long. Bracts 4 × 2 mm long .... **H.12. O. × symeonis**

11. Corollas 2.5 – 9.5 mm long. Calyx lower lip teeth much shorter than to ± as long as the upper lips, 0.1 – 1 mm long. Bracts c. 3 – 5 × 2 – 4 mm

12. Corollas 5 – 9.5 mm long. Calyces 3.5 – 4 mm long

13. Corollas pinkish. Bracts c. 5 × 4 mm, slightly purple........... .... **H.1. O. × adonidis**

13. Corollas white. Bracts c. 4 × 2.5 mm, green .. **H.3. O. × barbarae**

12. Corollas 2.5 – 6 mm long. Calyces c. 2 – 3.5 mm long

14. Spikes ovoid or (sub)globose, c. 4 × 4 mm. Calyces c. 2 mm long. Corollas pink .... **H.17. O. vulgare** ssp. **vulgare × Thymus species**

14. Spikes usually cylindrical, sometimes subglobose, up to 25 mm long, c. 5 – 6 mm wide. Calyces c. 2.5 – 3.5 mm long. Corollas usually white, sometimes pink

15. Bracts c. 4 × 3 mm. Calyces c. 3.5 mm long; lower lip teeth much shorter to ± as long as the upper lips .... **H.9. O. × majoricum**

15. Bracts c. 3 – 4 × 2 mm. Calyces c. 2.5 mm; lower lip teeth much to somewhat shorter than the upper lips

16. Stems more or less tomentellous; hairs c. 0.5 mm long. Leaves inconspicuously glandular punctate .... **H.2. O. × applii**

16. Stems more or less hirsute; hairs c. 1.5 mm long. Leaves obviously glandular punctate .... **H.6. O. × intercedens**

Descriptions

H.1. Origanum × adonidis Mouterde – Fig. 35.

O. × *adonidis* Mouterde (*O. libanoticum* × *syriacum* var. *bevanii*), Ann. Fac. Fr. Med. Beyrouth 6: 306 (1935); Davis, Kew Bull. 1949: 405 (1949). – Type: *Mouterde s.n.*, Lebanon, in valley of the Adonis (holo. G).

× *Majoranamaracus zernyi* Rechinger, Fedde Rep. 45: 95 (1938). – Type: *Zerny s.n.*, Lebanon, near monastery Becharré (holo. W).

Resembles *O. syriacum* var. *bevanii*, but bracts more or less purple and calyces 2-lipped for c. $\frac{2}{5}$. Stems up to 80 cm long, appressed pilosellous to glabrescent (hairs c. 0.5 mm long). *Leaves* more or less petiolate, c. 11 × 7 mm. *Spikes* subglobose or (longly) cylindrical, up to 30 mm long, c. 6 mm wide, (nearly) not nodding. *Bracts* c. 5 × 4 mm. *Flowers* 2 per verticillaster. *Calyces* ± tubular, c. 4 mm long; upper lips with small, ± deltoid teeth or lobes up to $\frac{1}{5}$, teeth up to 0.6 mm long; lower lips consisting of small ± deltoid teeth or lobes, much shorter than the upper lips, up to 1 mm long; throats pilose. *Corollas* 2-lipped for c. $\frac{1}{3}$, 5.5 – 9.5 mm long, pinkish. *Stamens* poorly developed and included or well developed and protruding; filaments up to 7 mm long.

Notes. *O.* × *adonidis* has been found in small numbers in several places in the Lebanon Mts., and is sometimes explicitly stated as found between the parents. 2. The toothing of the calyces and the length of the staminal filaments are variable in this hybrid.

LEBANON: Lebanon Mts., near monastery Becharré, c. 1350 m, *Zerny s.n.*, 2 July 1931 (W). Valley of Adonis, near Afka, 7 July 1932, *Mouterde s.n.* (type). Above Chaktoul, summer 1938, *Mouterde s.n.* (W). Above Baynu, c. 800 m, *Davis 6351* (K).

H.2. Origanum × applii (Domin) Boros – Fig. 35.

O. × *applii* (Domin) Boros (*O. majorana* × *vulgare* ssp. *vulgare*), Bot. Közlem. 35: 317 (1938); Appł, Preslia 6: 1 (1928). × *Origanomajorana applii* Domin, Preslia 13 – 15: 197 (1935). – Type: *Appl s.n.*, cultivated (holo. PRC).
O. paniculatum Koch, in Röhling & Koch, Deutschl. Fl. 4: 306 (1833). *Majorana paniculata* (Koch) Spenner, in Spenner & Nees von Esenbeck, Gen. Pl. Germ. 2: no. 15 (1843).
O. heracleoticum hort. ex Koch, in Röhling & Koch, Deutschl. Fl. 4: 306 (1833).

More or less intermediate between the parental species, with often cylindrical spikes, slightly purple bracts, pinkish corollas and 2-lipped calyces for c. $\frac{2}{5}$. Stems up to 50 cm long, more or less tomentellous (hairs c. 0.5 mm long). *Leaves* somewhat petiolate, c. 15 × 10 mm. *Spikes* subglobose to cylindrical, up to 25 mm long, c. 6 mm wide, not nodding. *Bracts* c. 4 × 2 mm, green or purplish. *Flowers* 2 per verticillaster. *Calyces* somewhat funnelshaped, c. 2.5 mm long; upper lips with ± deltoid teeth for $\frac{1}{3}$ – $\frac{1}{2}$, which are c. 0.5 mm long; lower lips somewhat shorter than the upper lips, consisting of ± triangular teeth, which are c. 0.8 mm long; throats pilosellous. *Corollas* 2-lipped for c. $\frac{1}{3}$, 4 – 6 mm long, white or pinkish. *Stamens* poorly developed and included or well developed and shortly protruding; filaments up to 4 mm long.

Notes. 1. *O.* × *applii* is since long known from gardens in West and Central Europe. 2. Appl gave a survey of the way in which several characters segregate in an F2 artificially made generation.

Cultivated in northern Chile, near Valparaiso, 1827, *unknown collector s.n.* (W). Cultivated, *Jacquin fil. s.n.* (W). Cultivated, 1926, *Appl s.n.* (type).

H.3. Origanum × barbarae Bornmüller – Fig. 35.

O. × *barbarae* Bornmüller (*O. ehrenbergii* × *syriacum* var. *bevanii*), Verh. Zool.-Bot. Geselsch. Wien 48: 615 (1898); Post & Dinsmore, Fl. Syr. Palest. Sin. 2: 334 (1933); Mouterde, Ann. Fac. Fr. Med. Beyrouth 6: 303 (1935); Rechinger, Fedde Rep. 45: 95 (1938); Rechinger, Arkiv Bot. 5: 385 (1963). – Type: *Bornmüller 1245*, Lebanon, Lebanon Mts. near Brummana (holo. B).

Like *O. ehrenbergii*, but calyces 2-lipped for c. $\frac{2}{5}$. Stems up to 60 cm long, more or less hirsute (hairs c. 1.5 mm long). *Leaves* more or less petiolate, c. 12 × 8 mm. *Spikes* usually cylindrical, up to 15 mm long, c. 5 mm wide, not nodding. *Bracts* c. 4 × 2.5 mm, green. Flowers 2 per verticillaster. *Calyces* somewhat funnel-shaped, c. 3.5 mm long; upper lips with ± deltoid teeth for c. $\frac{1}{2}$, which are c. 0.5 mm long; lower lips shorter than the upper lips, consisting of ± triangular teeth, which are c. 0.7 mm long; throats pilosellous. *Corollas* 2-lipped for c. $\frac{1}{4}$, 5 – 9 mm long, white. *Stamens* poorly developed, included; filaments up to 2 mm long.

Notes. 1. *O.* × *barbarae* has been found in a few locations in the Lebanon Mts., with both parental species nearby. 2. Mouterde (1935) stated that this hybrid seemed to multiply vigorously at one of these locations.

LEBANON: Lebanon Mts., near Brummana, c. 750 m, June 1897, *Bornmüller 1245* (type).

H.4. Origanum × dolichosiphon Davis – Fig. 35.

O. × *dolichosiphon* Davis (*O. amanum* × *laevigatum*), Kew Bull. 1951: 86 (1951). – Type: *Davis 16412*, Turkey, Seyhan, Dildil Dagh (holo. K, iso. E).

Like *O. amanum*, but spikes and flowers smaller. Stems up to 25 cm long, more or less hirtellous (hairs c. 0.3 mm long). *Leaves* (sub)sessile, c. 10 × 6 mm. *Spikes* cylindrical or subglobose, up to 25 mm long, c. 5 mm wide, not nodding. *Bracts* c. 7 × 2 mm, purple. *Flowers* 2 per verticillaster. *Calyces* tube-shaped, indistinctly 2-lipped for c. $\frac{1}{3}$, c. 6 mm long; upper lips for c. $\frac{4}{5}$ with triangular teeth, which are c. 2 mm long; lower lips as long as the upper lips, with triangular teeth, which are c. 2.3 mm long; throats pilose. *Corollas* 2-lipped for c. $\frac{1}{7}$, 11 – 18 mm long, pink. *Stamens* included; filaments up to 0.6 mm long.

Note. *O.* × *dolichosiphon* is only known from the type collection, when it was found in a small group of about six specimens, while the parental species were found growing in the immediate environs.

TURKEY. PROV. SEYHAN: Dildil Dagh, in steep gully between Başkonuş Y. and Huseyin Oluk Çe., 1800 m, 27 Aug. 1949, *Davis 16412* (type).

H.5. **Origanum × hybridinum** Miller – **Fig. 35.**

O. × *hybridinum* Miller (*O. dictamnus* × *sipyleum*) Gard. Dict. VIII Ed.: no. 12 (1768), pro sp. *Amaracus hybridus* (Miller) Jackson, Handl. R. B. G. Kew: 62 (1934); Lawrence, Baileya 2: 31 (1954); Wolf, Baileya 2: 60 (1954). – Type: *Miller s.n.*, cultivated (holo. BM).

O. pulchellum Boissier, Diagn. Pl. Or. Nov. 2(4): 9 (1859); Boissier, Fl. Or. 4: 547 (1879). *Amaracus pulchellus* (Boissier) Briquet, in Engler & Prantl, Nat. Pflanzenfam. 4(3a): 305 (1895); Wolf, Baileya 2: 60 (1954). – Type: *Boissier s.n.*, cultivated (holo. G).

Intermediate between the 2 parental species; often very floriferous hybrid like species in the section *Anatolicon;* calyces 1- or 2-lipped for c. $\frac{2}{5}$, with slightly or not developed teeth. Stems up to 50 cm long, more or less pilose (hairs c. 0.8 mm long). *Leaves* more or less petiolate, c. 12 × 8 mm. *Spikes* cylindrical to subglobose, up to 30 mm long, c. 11 mm wide, nodding. *Bracts* c. 7 × 5 mm, vividly purple. *Flowers* 2 per verticillaster. *Calyces* ± tubular, c. 4 mm long; upper lips (sub)entire; lower lips, when present, much shorter than the upper lips, consisting of small lobes, which are up to 0.5 mm long; throats pilosellous. *Corollas* 2-lipped for c. $\frac{1}{3}$, 8 – 12 mm long, pink, not saccate. *Stamens* well developed, protruding; filaments up to 10 mm long.

Notes. 1. Many authors, after Miller, have spelt the name of this hybrid *O. hybridum*. 2. *O.* × *hybridinum* is long known from gardens. Sometimes it has been stated as growing near the parental species. Although the stamens seem to be well developed, only a small percentage of the pollen is normally developed. It is questionable whether any fertile seeds are produced.

Cultivated in Browning's garden, Lincoln's Inn, England, *Miller s.n.* (type). Cultivated at Valeyres, *Boissier s.n.* (G). Hortus Berolinensis, 1868, *Watke s.n.* (JE). Hortus Kewensis, 1881, *Nicholson s.n.* (JE). Gardens near Willow Cottages, Kew, 15 Aug. 1914, *Fraser s.n.* (K). Garden of Gilling, England, 30 Aug. 1926, *Lester-Garland s.n.* (K). Blake's garden, Berkeley, California, 9 Sept. 1944, *Bracelin 1569* (L). Wisley Gardens, England, 5 Sept. 1952, *Boom 23168* (L).

H.6. **Origanum × intercedens** Rechinger – **Fig. 35.**

O. × *intercedens* Rechinger (*O. onites* × *vulgare* ssp. *hirtum*), Bot. Jahrb. 80: 395 (1961). – Type: *Rechinger 19403*, Greece, Euboea, near Paleochora (holo. W).

Resembles *O. vulgare* ssp. *hirtum*, but calyces 2-lipped for c. $\frac{2}{5}$, flowers somewhat flattened, and branching intermediate between the paniculate type of *O. vulgare* and the umbelliform type of *O. onites*. Stems up to 50 cm long, more or less hirsute (hairs c. 1.5 mm long). *Leaves* c. 12 × 8 mm, obviously glandular punctate, sometimes remotely serrulate. *Spikes* often cylindrical, up to 25 mm long, c. 5 mm wide, not nodding. *Bracts* c. 3 × 2 mm, green. *Flowers* 2 per verticillaster. *Calyces* more or less funnel-shaped, c. 2.5 mm long; upper lips for $\frac{2}{5}$ – $\frac{1}{2}$ with ± deltoid teeth, which are c. 0.5 mm long; lower lips (much) shorter than the upper lips, teeth nearly not developed to triangular and c. 1 mm long; throats pilosellous. *Corollas* 2-lipped for

c. $\frac{2}{5}$, 3 – 6 mm long, white. *Stamens* more or less developed, included or shortly protruding; filaments up to 3 mm long.

Note. Most probably *O.* × *intercedens* was collected in gardens long before it was found in nature, but those specimens were never recognized as being of hybrid origin. Until now it has been found only incidentally between herbarium specimens usually named *O. vulgare* ssp. *hirtum* (*O. hirtum*, *O. heracleoticum*).

Cultivated. *Burmann s.n.* (G). Cultivated, Hortus Landerhut, *Schultes s.n.* (L).
GREECE. EUBOEA: near Paleochora, on limestone, 29 June 1958, *Rechinger 19403* (type).
TURKEY. PROV. DENIZLI: Honaz Dağ, above Honaz, 24 Sept. 1973, *Tuzlaci 23532* (E).

H.7. Origanum × intermedium Davis – Fig. 35.

O. × *intermedium* Davis (*O. onites* × *sipyleum*), Kew Bull. 1949: 410 (1949). – Type: *Davis 13260*, Turkey, Denizli, near Denizli (holo. K, iso. E, para. G, W).

Intermediate between the parental species, with slender branches and small spikes with pink flowers, green bracts, and calyces, which are 1-lipped for c. $\frac{2}{5}$. Stems up to 50 cm long, more or less pilose (hairs c. 0.7 mm long). *Leaves* more or less petiolate, c. 8 × 4 mm. *Spikes* cylindrical, up to 16 mm long, c. 5 mm wide, not nodding. *Bracts* c. 5 × 3 mm. *Flowers* 2 per verticillaster. *Calyces* ± tubular, c. 3.5 mm long; upper lips subentire; throats pilosellous. *Corollas* 2-lipped for c. $\frac{1}{3}$, c. 8.5 mm long, pink. *Stamens* apparently well developed, protruding; filaments up to 6 mm long.

Note. *O.* × *intermedium* is only known from the type collection, one plant between the parental species.

TURKEY. PROV. DENIZLI: Taş Ocaği near Denizli, c. 600 m, 13 July 1947, *Davis 13260* (type).

H.8. Origanum × lirium Heldreich ex Halácsy – Fig. 35.

O. × *lirium* Heldreich ex Halácsy (*O. scabrum* × *vulgare* ssp. *hirtum*), Verh. Zool.-Bot. Geselsch. Wien 49: 192 (1899), *pro sp.*; Halácsy, Consp. Fl. Graec. 2: 554 (1902); Rechinger, Bot. Jahrb. 80: 395 (1961); Tutin et al., Fl. Eur. 3: 172 (1972). *Amaracus lirius* (Heldreich ex Halácsy) Hayek, Prodr. Fl. Penins. Balc. 2: 332 (1931); Rechinger, Fl. Aegaea: 531 (1943). *O. hybridum* Heldreich, Herb. Gr. Norm.: no. 783 (1858), *nomen nudum*. – Type: *Heldreich 783B*, Greece, Euboea, Mt. Dirphys (holo. W, iso. B, BM, G, JE, P, W, WU).

Somewhat like *O. scabrum*, but spikes much smaller and only slightly nodding. Stems up to 50 cm long, more or less hirtellous (hairs c. 0.4 mm long). *Leaves* (sub)sessile, c. 15 × 11 mm. *Spikes* subglobose to cylindrical, up to 30 mm long, c. 11 mm wide. *Bracts* c. 6 × 3 mm, partly purple. *Flowers* 2 per verticillaster. *Calyces* ± tubular, 2-lipped for c. $\frac{2}{5}$, c. 4 mm long; upper lips with ± triangular teeth for c.

$\frac{4}{5}$, which are 1 mm long; lower lips nearly as long as the upper lips, consisting of acute triangular teeth, which are c. 1.3 mm long; throats pilose. *Corollas* 2-lipped for c. $\frac{1}{3}$, 6 – 9 mm long, pink. *Stamens* well developed, protruding; filaments up to 6 mm long.

Notes. 1. Originally Heldreich (on herbarium label) conceived *O.* × *lirium* as a hybrid, later on as a "good" species. Comprehensive study of the three taxa in question showed that *O.* × *lirium* is intermediate between *O. scabrum* and *O. vulgare* ssp. *hirtum* in nearly all characters, and also that c. 50 percent of the pollen is poorly developed. Thus it seems right to consider this taxon as a hybrid which is "on the way to species". 2. On Mt. Delphi *O.* × *lirium* has been found in quantity, where it seems to form an independant population reproducing by seeds. Barnaby and Davis found a single specimen on Mt. Taygetos.

GREECE. EUBOEA: Mt. Delphi and Mt. Xerobuni, c. 1500 – 1700 m, 7 Aug. 1848, *Heldreich 783* (G, K, P, W). Near Liri, c. 1700 m, 1 Aug. 1871, *Orphanides 657* (W). In *Abies* zone of Mt. Dirphys near Elatakia, in neighbourhood of the spring Liri, c. 1300 m, 20 Aug. 1895, *Heldreich 783B* (type). On top of Mt. Delphi, 28 July 1901, *Leonis s.n.* (B, G. W). Mt. Delphi, in *Abies* region, Aug. 1910, *Tunta s.n.* (JE). Mt. Delphi, above the spring Liri, 24 July 1932, *Pinatzi 3530* (BM, G). Mt. Delphi, on calcareous rocks, c. 1000 – 1500 m, 13 – 17 July 1932, *Rechinger 2508* (B, K, W). MAINLAND: Peloponnisos, Mt. Taygetos, supra Tripi, on rocks in the subalpine region, c. 1700 m, Sept. 1938, *Barnaby & Davis s.n.* (K).

H.9. Origanum × majoricum Cambessedes – Fig. 36.

O. × *majoricum* Cambessedes (*O. majorana* × *vulgare* ssp. *virens*), Mem. Mus. Paris 14: 296 (1827), *pro sp.;* Willkomm & Lange, Prodr. Fl. Hisp. 2: 408 (1868); Tutin et al., Fl. Eur. 3: 172 (1972). *Majorana majorica* (Cambessedes) Briquet, in Engler & Prantl, Nat, Pflanzenfam. 4(3a): 308 (1895). *Amaracus majorica* (Cambessedes) Sampaio, Fl. Port: 511 (1947). – Type: *Cambessedes s.n.*, Spain, Mallorca, near Inca (holo. MPU, iso. K).
O. balearicum Pourret ex Lange, Videnskab. Medd. Kjöbenhavn 1863: 5 (1863).
O. lusitanicum Rouy, Naturaliste 12: 92 (1882). *O. majoricum* Cambessedes var. *lusitanicum* Rouy, Naturaliste 12: 92 (1882). *Majorana majorica* (Cambessedes) Briquet var. *lusitanica* (Rouy) Coutinho, Fl. Port.: 612 (1939).
O. paui Martinez, Mem. Soc. Esp. Hist. Nat. 14: 463 (1934).

Like *O. vulgare* ssp. *virens*, but spikes smaller, flowers somewhat flattened and calyces 2-lipped for c. $\frac{2}{5}$. Stems up to 60 cm long, tomentellous (hairs c. 0.5 mm long). *Leaves* petiolate, c. 9 × 5 mm. *Spikes* usually cylindrical, up to 20 mm long, c. 5 mm wide, not nodding. *Bracts* c. 4 × 3 mm, green. *Flowers* 2 per verticillaster. *Calyces* more or less funnel-shaped, c. 3.5 mm long; upper lips with ± deltoid teeth for $\frac{1}{3}$ – $\frac{1}{2}$, which are c. 0.4 mm long; lower lips much shorter than to nearly as long as the upper lips, sometimes subentire, usually consisting of triangular teeth, which are c. 0.7 mm long; throats pilosellous. *Corollas* 2-lipped for c. $\frac{1}{3}$, 2.5 – 6 mm long, white. *Stamens* poorly developed to apparently well developed; filaments up to 2.5 mm long.

N o t e. *O.* × *majoricum* is known from several natural sites on the Balearic Islands, and in Spain and Portugal. This is not surprising because *O. majorana* occurs subspontaneously and *O. vulgare* ssp. *virens* is very common in several parts of these countries. The hybrid is also found in gardens.

SPAIN. MALLORCA: near Inca, *Cambessedes s.n.* (type). MAINLAND: Alicante, cultivated near Biar, *Sennen 9962* (W).

PORTUGAL: near Tapada do Alfeite, *Welwitsch s.n.* (COI). Cultivated in Blake's Garden, Berkeley, California, 2 Aug. 1942, *Bracelin 2225* (L).

H.10. Origanum × minoanum Davis – Fig. 36.

O. × *minoanum* Davis (*O. microphyllum* × *vulgare* ssp. *hirtum*), Not. R. B. G. Edinburgh 21: 137 (1953). *Majorana leptoclados* Rechinger, Denkschr. Akad. Wiss. Wien, Math..Nat. Kl. 105(2): 125 (1943). – Type: *Rechinger 13551*, Greece, Kriti, between Paleochora and Vutas (holo. W, iso. G).

Like *O. microphyllum*, but in most parts stronger, and calyces 2-lipped and slightly dentate. Stems up to 45 cm long, appressed pilose (hairs c. 0.8 mm long). *Leaves* more or less petiolate, c. 7 × 4 mm. *Spikes* subglobose, up to 12 mm long, c. 4 mm wide, not nodding. *Bracts* c. 3 × 2 mm, green. *Flowers* 2 per verticillaster. *Calyces* ± tubular, 2-lipped for c. $\frac{1}{5}$, c. 2.2 mm long; upper lips (sub)entire or with very small, ± deltoid teeth, which are c. 0.2 mm long; lower lips slightly shorter than the upper lips, consisting of ± deltoid teeth, which are c. 0.2 mm long; throats pilosellous. *Corollas* 2-lipped for c. $\frac{2}{5}$, somewhat flattened, c. 5.5 mm long, pinkish. *Stamens* slightly or not developed, included to slightly protruding; filaments up to 2 mm long.

N o t e. *O.* × *minoanum* has been found twice in nature, in few specimens together. In both cases only *O. vulgare* ssp. *hirtum* was found growing nearby.

GREECE. KRITI: Selinos, on schistous rocks in valley of Pelekaniotikos, between Palaeochora and Vutas, 2 June 1942, *Rechinger 13551* (type). Above Theriso, 3 Aug. 1950, *Davis 18149* (E).

H.11. Origanum × pabotii Mouterde – Fig. 36.

O. × *pabotii* Mouterde (*O. bargyli* × *syriacum* var. *bevanii*), Saussurea 4: 22 (1973). – Type: *Pabot s.n.*, Syria, w. of Alaouite Mts. (holo. & iso. G).

Somewhat like *O. syriacum* var. *bevanii*, but bracts slightly purple and calyces 2-lipped for c. $\frac{2}{5}$. Stems up to 45 cm long, more or less (appressed) pilose(llous), (hairs c. 0.5 mm long). *Leaves* more or less petiolate, c. 12 × 8 mm. *Spikes* cylindrical, up to 25 mm long, c. 5 mm wide, not nodding. *Bracts* c. 5 × 3.5 mm. *Flowers* 2 per verticillaster. *Calyces* ± tubular, c. 4 mm long; upper lips subentire or with very small teeth, which are c. 0.1 mm long; lower lips shorter than the upper lips, consisting of ± deltoid teeth, which are c. 0.7 mm long; throats pilose. *Corollas* 2-lipped for c. $\frac{1}{5}$, 7 – 9 mm long, pinkish. *Stamens* more or less developed, usually shortly protruding; filaments up to 3 mm long.

Note. *O.* × *pabotii* is only known from the type specimens, which were found growing with both parental species.

SYRIA: w. of Alaouite Mts., c. 1100 m, 25 July 1955, *Pabot s.n.* (type).

H.12. Origanum × symeonis Mouterde – Fig. 36.

O. × *symeonis* Mouterde (*O. laevigatum* × *syriacum* var. *bevanii*), Ann. Fac. Fr. Med. Beyrouth 6: 305 (1935), (French diagnosis), Saussurea 4: 23 (1973), (Latin diagnosis). – Type: *Mouterde 3390*, Turkey, between Antioche and Soueidié (holo. G).

O. haradjanii Rechinger, Österr. Bot. Zeitsch. 99: 64 (1952); Rechinger, Arkiv Bot. 5: 385 (1963). – Type: *Haradjian 3175*, Turkey, Hatay, Djebel Mousa near Antioche (holo. G, iso. W).

Somewhat like *O. laevigatum*, but bracts green or sometimes slightly purple, and calyces 2-lipped for c. $\frac{2}{5}$. Stems up to 70 cm long, more or less tomentose (hairs c. 0.7 mm long). *Leaves* petiolate, c. 11 × 5 mm. *Spikes* usually cylindrical, up to 16 mm long, c. 4 mm wide, not nodding. *Bracts* c. 4 × 2 mm. *Flowers* 2 per verticillaster. *Calyces* tubular, c. 4 mm long; upper lips with deltoid teeth for c. $\frac{1}{3}$, which are c. 0.5 mm long; lower lips somewhat shorter than to nearly as long as the upper lips, consisting of triangular teeth, which are c. 1 mm long; throats pilose. *Corollas* 2-lipped for c. $\frac{1}{5}$, 9 – 12 mm long, pinkish. *Stamens* more or less developed, (sub)-included or shortly protruding; filaments up to 4.5 mm long.

Note. *Origanum* × *symeonis* was found by Mouterde in rather large numbers in an area where both parental species were mixed with each other.

TURKEY. PROV. HATAY: Amanus, Djebel Mousa near Antioche, July 1909, *Haradjian 3175* (G, W). Jabal Sema'ane between Antioche and Soueidié, Aug. 1934, *Mouterde 3390* (type).

H.13. Origanum amanum × dictamnus – Fig. 36.

Much like *O. dictamnus*, but calyces 2-lipped for c. $\frac{1}{2}$ with clearly dentate lower lips, and corollas larger with reduced, included stamens. Stems c. 25 cm long, pilose (hairs c. 1 mm long). *Leaves* subsessile, c. 16 × 14 mm. *Spikes* subglobose, c. 15 × 15 mm, nodding. *Bracts* c. 10 × 8 mm, partly purple. *Flowers* 2 per verticillaster. *Calyces* ± tubular, c. 6 mm long; upper lips entire; lower lips much shorter than the upper lips, consisting of slightly triangular teeth, which are c. 1 mm long; throats pilose. *Corollas* 2-lipped for c. $\frac{1}{5}$, 15 – 20 mm long, purplish. *Stamens* not well developed, included; filaments up to 3 mm long.

Note. *O. amanum* × *dictamnus* is an artificial hybrid, made with garden grown parents.

Cultivated in botanical garden of Jena, Aug. 1977, *Meyer s.n.* (JE).

H.14. Origanum calcaratum × dictamnus – Fig. 36.

O. calcaratum × *dictamnus;* Davis, Not. R. B. G. Edinburgh 21: 137 (1953).

Not studied by the author. The 2 parental species are rather closely related, so some characters of the hybrid are easily to conceive. Stems c. 30 cm long, more or less lanate. *Leaves* more or less petiolate, c. 16 × 15 mm. *Spikes* ovoid, c. 20 × 14 mm, nodding. *Bracts* c. 10 × 6 mm, partly purple. *Calyces* 1-lipped for c. $\frac{3}{5}$, c. 6 mm long; upper lips (sub)entire. *Corollas* 2-lipped for c. $\frac{1}{3}$, c. 12 mm long, pink, saccate.

Note. Davis reported this hybrid from garden environments. In nature it may be expected on eastern Kriti, where the 2 parental species are able to meet. It will be by no means easy to discern this hybrid from the parental species.

H.15. Origanum micranthum × vulgare ssp. hirtum – Fig. 36.

O. micranthum × *vulgare* ssp. *hirtum;* Davis, Kew Bull. 1951: 88 (1951).

Like *O. micranthum*, but with subregularly 5 toothed calyces for c. $\frac{1}{3}$. Stems up to 40 cm, more or less piloso-tomentose (hairs c. 1 mm long). *Leaves* more or less petiolate, c. 10 × 7 mm. *Spikes* subglobose or cylindrical, up to 10 mm long, c. 4 mm wide, not nodding. *Bracts* c. 3.5 × 2.5 mm, greyish. *Flowers* 2 per verticillaster. *Calyces* ± tubular, c. 2 mm long; teeth c. 0.7 mm long; throats pilosellous. *Corollas* 2-lipped for c. $\frac{1}{4}$, c. 3.5 mm long, white. *Stamens* almost undeveloped, included; filaments c. 0.5 mm long.

Note. This hybrid which is more or less putative has been found twice. Balansa collected it in 1855 with specimens of *O. micranthum*, on the only site where this species occurs (Cilician Gates in southern Turkey). In 1949 Davis could not detect this species but only hybrid forms and *O. vulgare* ssp. *hirtum*. Further investigations are highly desirable.

TURKEY. PROV. IÇEL: Cilician Gates, rocks near ruins of castle, Aug. 1855, *Balansa 537* (p.p.) (G). PROV. SEYHAN: Bulgar Dağ, between Pozanti and Meydan Y., 1500 m, 1 Sept. 1949, *Davis 16586* (E).

H.16. Origanum sipyleum × vulgare ssp. hirtum – Fig. 36.

Like *O. sipyleum*, but leaves more glandular punctate, and calyces 2-lipped for c. $\frac{1}{2}$. Stems up to 40 cm long, more or less pilosellous (hairs c. 0.5 mm long). *Leaves* c. 9 × 5 mm. *Spikes* ovoid, up to 14 mm long, c. 7 mm wide, more or less nodding. *Bracts* c. 5 × 2.5 mm, purplish. *Flowers* 2 per verticillaster. *Calyces* ± tubular, c. 3 mm long; upper lips with small ± deltoid teeth for c. $\frac{1}{6}$, which are c. 0.2 mm long; lower lips much shorter than the upper lips, consisting of ± deltoid teeth, which are c. 0.3 mm long; throats pilosellous. *Corollas* 2-lipped for c. $\frac{1}{3}$, c. 6 mm long, pink. *Stamens* poorly developed, included; filaments up to 2 mm long.

Note. *O. sipyleum* × *vulgare* ssp. *hirtum* is putative and has been found once as a herbarium specimen named *O. sipyleum*.

TURKEY. PROV. IZMIR: near Izmir, 1812, *Rennard s.n.* (BM).

H.17. Origanum vulgare ssp. vulgare × Thymus species – Fig. 36.

Most like *O. vulgare* ssp. *vulgare*, but with more and slender branches, smaller spikes, and 2-lipped calyces for c. $\frac{2}{5}$. Stems up to 45 cm long, more or less appressed pilose (hairs c. 1 mm long). *Leaves* more or less pilose, c. 11 × 7 mm. *Spikes* ovoid or (sub)globose, c. 4 × 4 mm, crowded, not nodding. *Bracts* c. 3 × 2 mm, green or slightly purple. *Flowers* 2 per verticillaster. *Calyces* somewhat funnel-shaped, c. 2 mm long; upper lips with deltoid teeth for c. $\frac{1}{2}$, which are c. 0.3 mm long; lower lips nearly as long as the upper lips, consisting of ± triangular teeth, which are c. 0.8 mm long; throats pilosellous. *Corollas* 2-lipped for c. $\frac{2}{5}$, c. 5 mm long, pink. *Stamens* apparently well developed, shortly protruding to included; filaments up to 3 mm long.

Note. *O. vulgare* ssp. *vulgare* × *Thymus* species is highly putative. It was found between herbarium specimens of *O. vulgare*. The site where it was collected is rather well known. Further investigation is much needed.

AUSTRIA: near Hard at the Bodensee, 4 July 1937, *Schneider s.n.* (G, W).

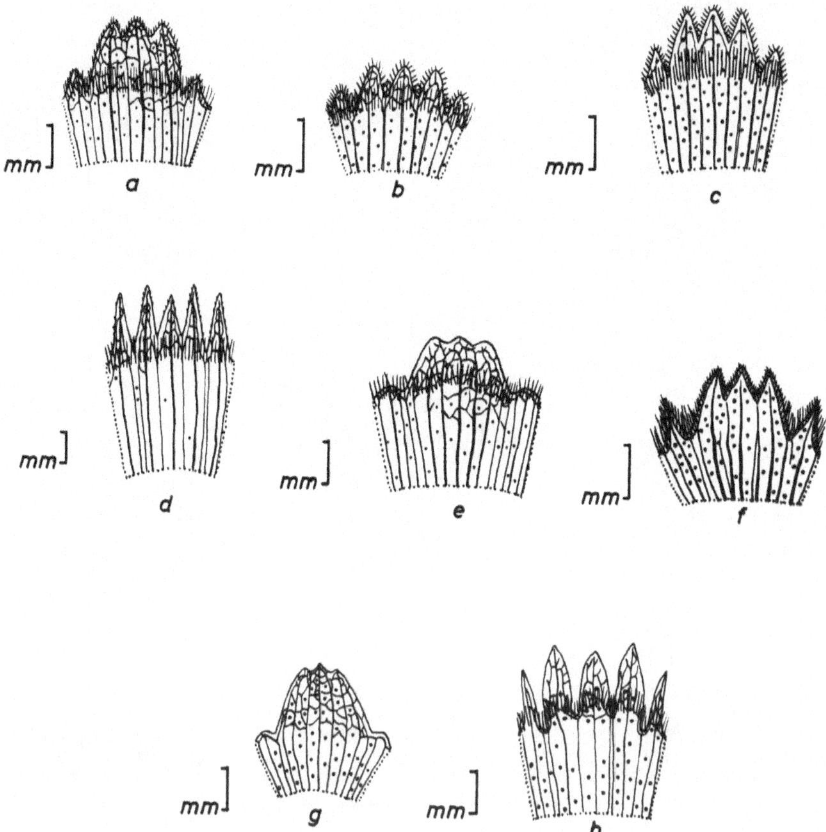

Figure 35. Calyces of hybrids cut through the lower lip: a. *O.* × *adonidis;* b. *O.* × *applii;* c. *O.* × *barbarae;* d. *O.* × *dolichosiphon;* e. *O.* × *hybridinum;* f. *O.* × *intercedens;* g. *O.* × *intermedium;* h. *O.* × *lirium.*

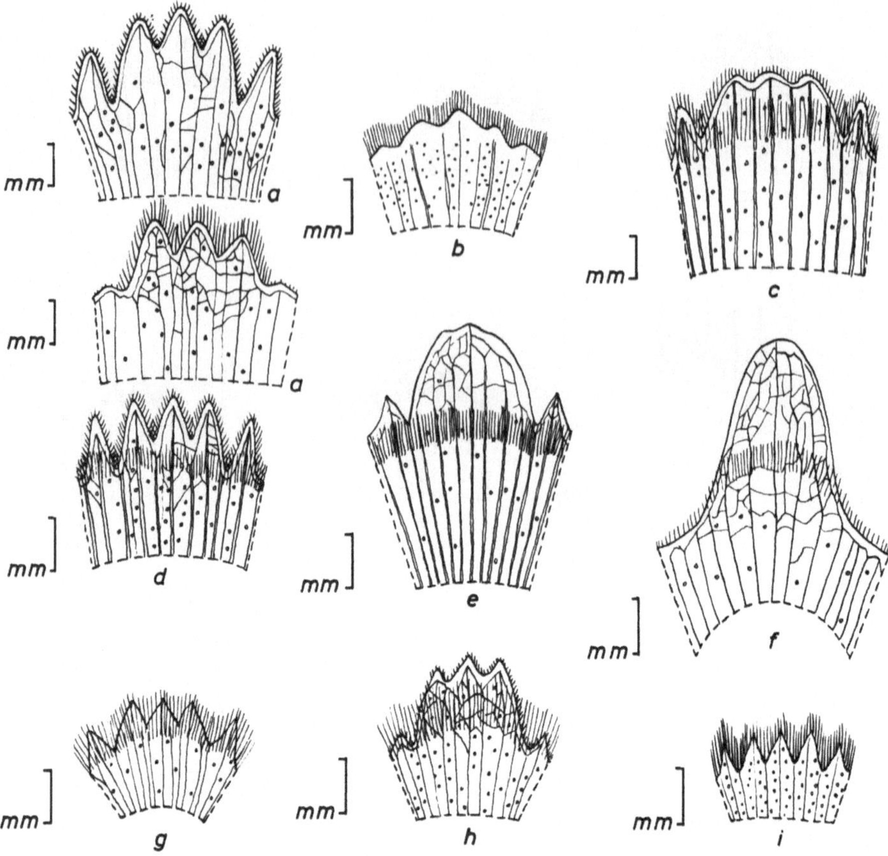

Figure 36. Calyces of hybrids cut through the lower lip: a. *O.* × *majoricum;* b. *O.* × *minoanum;* c. *O.* × *pabotii;* d. *O.* × *symeonis;* e. *O. amanum* × *dictamnus;* f. *O. calcaratum* × *dictamnus;* g. *O. micranthum* × *vulgare* ssp. *hirtum;* h. *O. sipyleum* × *vulgare* ssp. *hirtum;* i. *O. vulgare* ssp. *vulgare* × *Thymus* species.

II.6. NOMINA DUBIA

Majorana neglecta (Vogel) Walpers, Rep. Bot. Syst. 3: 697 (1844). Synonym of *O. neglectum* (listed below).
Majorana turbinata (Vogel) Walpers, Rep. Bot. Syst. 3: 697 (1844). Synonym of *O. turbinatum* (listed below).
Origanum acinacifolium Wallroth ex Steudel, Nomencl. Bot. 2: 226 (1841). Misspelling of *O. acinifolium*.
Origanum acinifolium Wallroth ex Bentham, Lab. Gen. Sp.: 728 (1834). Synonym of *O.* × *applii* or *O.* × *majoricum*.
Origanum amaracus Bedevian (nomen nudum), Ill. Polygyl. Dict. Pl. Names: 427 (1936). Synonym of *O. dictamnus?*
Origanum candelabrum Lojacono Pojero, Fl. Sicula 2: 196 (1904). Synonym of *O. vulgare* ssp. *hirtum?*
Origanum creticum Loureiro ex Kosteletzky, Med.-Pharm. Fl. 3: 768 (1834). Synonym of *O. loureiri* (listed below).
Origanum dilatatum Klokov in Fl. R.S.S. Ucr. 9: 664 (1960). Synonym of one of the ssp. of *O. vulgare*.
Origanum foliosum Lojacono Pojero, Fl. Sicula 2: 196 (1904). Synonym of *O. vulgare* ssp. *hirtum?*
Origanum fortuitum Savi, Osserv. Gen. Origanum: 14 (1840). *O. majorana* × *vulgare* ssp. *viride?*
Origanum heracleoticum Loureiro ex Kosteletzky, Med.-Pharm. Fl. 3: 768 (1831). Synonym of *O. loureiri* (listed below).
Origanum hereaclontica Rafinesque, Fl. Tell. 3: 86 (1836). Most probably misspelling of *O. heracleoticum*, so probably synonym of *O. vulgare* ssp. *viride*.
Origanum laxiflorum Royle ex Bentham, Bot. Miscell. 3: 376 (1833). Synonym of one of the ssp. of *O. vulgare*.
Origanum loureiri Kosteletzky, Med.-Pharm. Fl. 3: 768 (1834). Probably synonym of one of the ssp. of *O. vulgare*.
Origanum micranthum hort. ex Colla, Hort. Ripul. 4: 53 (1832). Synonym of one of the ssp. of *O. vulgare?*
Origanum nebrodense Tineo ex Lojacono Pojero, Fl. Sicula 2: 196 (1904). *O. majorana* × *vulgare* ssp. *viride?*
Origanum neglectum Vogel, Linnaea 15: 81 (1841). Most probably synonym of *O.* × *intercedens*.
Origanum nummulariaefolia Holmes, Perf. Ess. Oil. Rec. 4: 69 (1913). Synonym of *O. majorana?*
Origanum salvifolium Roth, Cat. Bot. 2: 48 (1800). Synonym of *O. majorana* or *O.* × *applii?*
Origanum serpylliforme Fischer et Meyer, Index Sem. Hort. Petropolitanus 11: 59 (1835). Synonym of one of the ssp. of *O. vulgare?*
Origanum turbinatum Vogel, Linnaea 15: 77 (1841). Possibly synonym of *O.* × *applii*.
Origanum vulgare Linnaeus var. *laxiflorum* (Royle ex Bentham) Briquet, in Engler & Prantl, Nat. Pflanzenfam. 4(3a), 309 (1895). Synonym of *O. laxiflorum* Royle ex Bentham (listed above).
Oroga hereaclontica Rafinesque, Fl. Tell. 3: 86 (1836). Synonym of *O. hereaclontica* Rafinesque (listed above).

II.7. SPECIES EXCLUDENDA

As mentioned in I.2. Kuntze once brought all *Thymus* species under the genus *Origanum*. These are not mentioned below, except two.

Majorana aegyptiaca (Linnaeus) Kosteletzky, Med.-Pharm. Fl. 3: 770 (1834). Synonym of *Origanum aegyptiacum* (listed below).
Majorana rotundifolia Miller, Gard. Dict. IV Ed.: no. 2 (1754).
Origanum aegyptiacum Linnaeus, Sp. Pl.: 588 (1753). What Linnaeus understood by this species is uncertain, probably a *Pogostemon* or *Plectranthus* species.
Origanum bengalense Heyne ex Wallich, Num. List: no. 1532 (1829). Synonym of *Pogostemon patchouly* Pellet.
Origanum benghalense Burman, Fl. Indica: 128 (1768). Synonym of *Pogostemon plectranthoides* Desf.
Origanum clinopodioides Walter, Fl. Caroliniana: 165 (1788). Synonym of *Pycnanthemum setosum* Nutt.

Origanum elegans Stephan ex Steudel, Nomencl. Bot. 2: 227 (1841). Synonym of *Dracocephalum origanoides* Steph. ex Willd.
Origanum flexuosum Walter, Fl. Caroliniana: 165 (1788). Synonym of *Pycnanthemum flexusosum* (Walt.) B.S.P.
Origanum incanum Walter, Fl. Caroliniana: 165 (1788). Synonym of *Pycnanthemum setosum* Nutt.
Origanum indicum Roth, Nov. Pl. Sp.: 265 (1821). Synonym of *Pogostemon plectranthoides* Desf.
Origanum lanatum Bojer ex Nees, in de Candolle, Prodr. Syst. Nat. 11: 721 (1847). Synonym of *Nelsonia campestris* R. Br.
Origanum punctatum Poiret, in Lamarck, Encycl. Méth. Bot. IVème Suppl.: 186 (1816). Synonym of *Pycnanthemum incanum* (L.) Michx.
Origanum richardii (Persoon) Kuntze, Rev. Gen. Pl. 2: 528 (1891). Synonym of *Thymus richardii* (Persoon) and not of *O.* × *majoricum* (e.g. Nyman, Consp. Fl. Eur.: 592 (1881).
Origanum thymus Kuntze, Rev. Gen. Pl. 2: 528 (1891). Synonym of *Thymus vulgaris* L.
Origanum vulgare Müller, Ic. Pl. Daniae 4: table 638 (1777). Synonym of *Mentha aquatica* L.
Origanum wateriense Roxburgh ex Hooker, Fl. Brit. India 4: 627 (1885). Synonym of *Anisochilus carnosus* Wall.

II.8. INDEX OF COLLECTIONS

Each collector's name is followed by his collection number(s) and the taxon number(s) (separated by a colon). In the taxon number(s) a etc. refers to subspecific taxa and H to hybrids. Unnumbered collections are not included.

Aellen 799: 34f — Agnew 752: 34c — Alice N2-844: 34d — Alston & Sandwith 2236: 34d — Amdursky: see Zohary — Anders 5240: 34c; 11096: 34f — Archibald 2321: 34a — Asplund 1360: 34a — Assadi: see Wendelbo — Aucher 1656: 3; 1656bis: 10.

Balansa (1854) 317: 26; (1854) 319: 34d; (1854) 328: 12; (1855) 537 (pro parte): H. 15; (1855) 537 (pro parte): 22; (1855) 542: 1; (1857) 1174: 12; (1866) 1523: 34f — Bailey 811: 34a; 961: 34a; 1233B: 34a — Baldacci 41: 26; 85: 23; 91: 23; 165: 26; 190: 3; 345: 23 — Balls 1707: 14; 1721: 34f — Bally 13162: 34d — Bangerter see Melderis — de Barros e Cunha 659A: 34e — Battandier & Trabut 364: 34b; 563: 32 — Bauer 105: 34d — Bauer & Spitzenberger 1153: 34d — Bayer 90: 34a — Baytop 1403: 34d: 6031: 34d: 13360: 34a; 13535: 34d; 13648: 34d; 14138A: 34a — Beguinot & Vaccari 180: 8 — Beug: see Wagnitz — Bianor 1427: 34e; 2900: 34b — Billot 3451: 34a — Blanche 86: 27b — Blumberg 986: 34a — Boom 23168: H. 5 — van Borssum Waalkes 5147: 34a — Boulos 4469: 7 — Bourgeau 39: 34b; 139: 26; 219: 26; 264: 34e; 700: 34f; 1426: 34e; 1995: 34e — Bornmüller (see also Sintenis) (1912) B1023: 28; (1900) 1067: 34e; (1900) 1068: 34e; (1897) 1245: H. 3; (1897) 1247: 27a; (1897) 1248: 27b; (1897) 1249: 36; (1889) 1432: 34f; (1917) 1834: 34d; (1893) 3481: 34c; (1899) 5463: 12; (1906) 9855: 26; (1910) 12321: 10; (1910) 12323: 36 — Bracelin 1569: H. 5; 2225: H. 9 — Briquet 1466: 34b — Burchard 48: 34e — Burdet 15: 27a; 60: 27a; 328: 27a — Buttler 14332: 34f; 15529: 12; 15567: 34a — Buttler & Erben 17345: 26; 17513: 26 — Buysman 119: 34a.

Chapman 348: 3 — Chien 1294: 34f — Choroschkov 3785A: 34a — Choulette 79: 34b — Cobham: see Wendelbo — Corstorphine 1326: 34a — Crowfoot 84: 27a.

Danin 13141: 30 — Darrah 338: 34d — Davis (see also Davis & Hedge, Davis & Polunin and Heywood) 944: 2; 3388: 3; 6315: 10; 6351: H. 1; 6439A: 27b; 13230: 34d; 13256: 12; 13260: H. 7; 13401: 9; 13589: 34d; 13599: 9; 13636: 34d; 14078: 6; 14099: 6; 14185: 24; 14276: 5; 14397: 5; 14439: 5; 14720: 21; 15773: 24; 15857: 5; 15893: 34f; 15930: 34f; 16194: 18; 16333A: 25; 16371: 37; 16390: 20; 16412: H. 4; 16439: 20; 16586: H. 15; 16588: 34d; 18005: 13; 18085: 26; 18149: H. 10; 18349: 34d; 18418: 34d; 20237: 34c; 20773: 34f; 20839: 34f; 20990: 34a; 23076: 34c; 23447: 34c; 23932: 34c; 25013: 34f; 30037: 34f; 31441: 34c; 32041: 34f; 32413: 34f; 32497: 34a; 33886: 34a; 36327: 34f; 37633: 34f; 37911: 34f; 38430: 34f; 39172A: 34d; 45325: 14; 46051: 34c; 46451: 14; 46579: 34a; 47669: 19 — Davis & Hedge 29854: 19; 30051B: 19; 30953: 14; 31067: 14; 31637: 14; 31771: 14 — Davis & Polunin 24283: 14 — Daveau 651: 34e — Demirdögen 2580: 24 — Devos: see Thielens — Dinsmore (see also Meyers) 13929: 10 — Dominguez: see Gibbs — Dörfler 663: 4; 880: 4; 1045: 4; 1086: 23; 1089: 4; 4755: 4 — Dubois & Faurel 1005: 32 — Dumansky 378C: 34a — Durando 19: 34b — Duval-Jouve 1652: 32 — Duyfjes & Kanis 394: 34a.

Ecoyomides 1190: 25; 1202: 25 — Edelberg 1198: 34c; 1393: 34c — Edmondson 715: 34f — Erben: see Buttler.

Fan & Li 173: 34f — Faurel: see Dubois — Federov: see Gomolitzky — Fernandes, Fernandes & Matos 1748: 34e — Fernandes & Matos 348A: 34e — Fernandes & Soeesa 2649: 34e — Findlay 240: 34d — Fokkinga 152: 11; 164: 11; 166: 11; 167: 11; 171: 11; 179: 11; 182: 11; 185: 11; 201: 11; 214: 11; 223: 11; 241: 11; 246: 11; 255: 11; 258: 11 — Font Quer 351: 35; 352: 33; 551: 31; 577: 33; 578: 33 — Forrest 20372: 34a — Forsyth Major 203: 26; 262: 13; 713: 12; 726: 34d — Furse 2877: 34a; 9144: 19.

Gaillardot 521: 36 — Gandoger 1055: 34f; 7649: 34d; 12388: 34d; 12668: 34d — Garcia 705: 34e — Gattefossé 983: 35 — Gibbs & Dominguez D224: 34e — Gleisberg 205: 14 — Gomolitzky & Federov 272: 34c — Görz S12: 34f — Grant 16309: 34f; 16379: 34f; 16489: 34f; 16499: 34f — Greuter S3709: 23; S3757: 34d; S4704: 34d; 7536: 4; 7627: 4; 7644: 2; 7688: 4; 7795: 4 — Guiol 268: 26; 1393: 34d; 2119: 23; 2120: 4; 2534: 11; 2549: 2; 2563: 34f.

Halácsy: see Leonis — Handel-Mazzetti (see also Pichler) 198: 34f; 1936: 34c; 2892: 34c; 2913: 14 — Haradjian 451: 27b; 491: 37; 668: 27b; 886: 25; 954: 3; 973: 25; 2433: 20; 2449: 27b; 2621: 37; 3175: H. 12; 3200: 37; 3884: 20; 3904: 37; 4644: 27b; 4645: 37 — Haussknecht 7003: 34f — Hedge: see Davis — Heldreich 783: H. 8; 783B: H. 8; 784: 11 — Henderson: see Stainton — Heywood 1523: 34e — Heywood & Davis 183: 34e — Holmberg 1887: 19; 2227: 19 — Holmboe 798: 25 — Hoogland 1947-269: 34a — Huber-Morath 9394: 17; 11382: 34c; 12707: 12; 12708: 27b; 14167: 34f; 15266: 15; 15267: 14; 15268: 19; 15553: 34f; 15554: 34f; 17209: 12.

Jacobs 811: 34a — Jahandiez 633: 35; 911: 34e — Jamin 74: 34b; 199: 32.

Kanis: see Duyfjes — Kennedy 742: 25; 1633: 3 — Khan et al. 683: 34f — Koelz 12926: 34c; 16155: 34f; 16976: 34f — Kotschy (1862) 48: 27b; (1853) 238: 1; (1855) 256: 10; (1853) 256A: 34d; (1855) 260: 27b; (1853) 261: 22; (1853) 355C: 34d; (1859) 444: 34c; (1836) 471: 22; (1836) 472: 34d; (1862) 937: 25 — Kralik 290: 34b.

Lamond 4837: 34a; 2065: 34c — Lemperg 511: 23 — Leonis & Halácsy 230: 23 — Letourneux 355: 26 — Levier: see Sommier — Li: see Fan — Licent 1502: 34f.

Matos: see Fernandes — Maurico: see Sennen — McNeill 603: 34c — Meinke 113: 16 — Melderis & Bangerter 129: 34a — Merton 1944: 3 — Meyers B1023: 28 — Meyers & Dinsmore B4037: 27a — Mouterde 3390: H. 12 — Munby 69: 34b.

Neubauer 988: 34c; 4404: 34c; 5004: 34c — Noei 209: 34f.

Oberneder 6633: 34d — Orphanides 42: 11; 143: 26; 657: H. 8; 969: 11; 1168: 2.

Pampanini & Pichi-Sermolli 6704: 8; 6705: 8; 6706: 8; 6707: 8; 6708: 8; 6709: 8 — Paris 370: 34b — Pau 1363: 34a — Peşmen 1040: 34c — Petter 266: 34d — Peyron 689: 36 — Phitos 3196: 34d; 3938: 11; 4030: 34d; 5388: 34d — Pichi-Sermolli: see Pampanini — Pichler 526: 13 — Pichler & Handel-Mazzetti 181: 34d — Pinatzi 3527: 11; 3530: H. 8; 14024: 23 — Pitard 1625: 35; 1626: 35 — Podlech 11881: 34c Polunin: see Davis — Porta & Rigo 576: 34e — Post 244: 28; 323: 20 — Pravitz 947: 34f.

Radermacher 431: 34d; 475: 34d; 547: 34d — Rahman 152: 34f — Reboud 1791: 34b — Rechinger (see also Rechinger & Sleumer) 1014: 26; 1054: 26; 1212: 26; 1874: 34f; 1919: 26; 2018: 34f; 2023: 12; 2069: 34f; 2314: 2; 2363: 2; 2508: H. 8; 2513: 11; 2519: 34d; 2572: 11; 5564: 34f; 6586: 34f; 8187: 13; 8255: 26; 13024: 26; 13315: 4; 13386: 34d; 13531: 4; 13551: H. 10; 13554: 34d; 13685: 23; 13741: 23; 13948: 26; 14373: 23; 14429: 34d; 15048: 34d; 15788: 34d; 16720: 11; 16905: 34d; 18181: 34d; 18194: 34d; 18242: 11; 18711: 34d; 19403: H. 6; 19465: 34f; 20035: 34d; 20077: 26; 20606: 34d; 21141: 34f; 30454: 34f; 31325: 34c; 31629: 34c — Rechinger & Sleumer 1212: 34a — van Regteren Altena 1990: 34a — Reis Moura 727: 34e — Rigo (see also Porta) 343: 34f; 4291: 34f — Reverchon 77: 34b; 128: 23; 134: 34d; 538: 25 — Ross 64: 26; 663: 34f — Rothmaler 13471: 34e; 14027: 34e; 14092: 34e.

Salle 67: 34b — Sandwith (see also Alston) 1940: 34a; 2423: 8 — Schindler 160: 34f — Schlagintweit 3589: 34f — Schneider 441: 34a — Schousbou 107: 35 — Schultz 1444: 34e — Schwarz 443: 9; 792: 12; 965: 12 — Sennen 2046: 34a; 3535: 34a; 3536: 34a; 9962: H. 9 — Sennen & Maurico 7985: 25; 8480: 31; 8931: 31 — Siehe 330: 1; 669: 22 — da Silva 1103: 34e — Sinclair 4651: 34e — Sintenis (1883) 456: 34d; (1883) 456B: 34d; (1889) 1027: 14; (1890) 2940: 34c; (1890) 2996: 14; (1890) 3252: 14; (1892) 4911: 12; (1894) 6102: 14 — Sintenis & Bormüller (1891) 1404: 34d — Sleumer: see Rechinger — Smith 4785: 34a — Soeesa: see Fernandes — Sommier & Levier 1064: 19 — Sorger 63 - 25 - 20: 26; 65 - 33 - 65: 26; 66 - 13 - 7: 25; 66 - 30 - 8: 25; 66 - 48 - 9: 26; 67 - 21 - 5: 26; 67 - 44 - 12: 12; 68 - 43 - 49: 34f; 68 - 46 - 40: 12; 68 - 50 - 9: 34f; 69 - 13 - 59: 34a; 69 - 21 - 1A: 34f; 69 - 23 - 31: 34f; 69 - 37 - 31: 14; 71 - 13 - 17: 26; 71 - 14 - 5: 25; 71 - 27 - 5: 26; 71 - 31 - 28: 27b — Spicer 92: 11 — Spitzenberger: see Bauer — Stainton 7747: 34f — Stainton & Henderson 5950: 19 — Strid 843: 34d.

Tacik 571: 34a — Termé 14472E: 34f — Thielens & Devos 320: 34a — Tobey 1179: 34f; 1328: 34f; 2178: 34f — Tong 13: 34c; 34: 34c — Trabut: see Battandier — Tuzlaci 23532: H. 6.

Vaccari see Beguinot.

Wagenitz & Beug 243: 34f — Walas 358: 34a — Wallich 1564: 34f — Walton 64: 34f; 141: 34f — Welwitsch 150: 34e — Wendelbo 1499: 34f; 9242: 34c; 9719: 34c — Wendelbo & Assadi 14585: 34f — Wendelbo & Cobham 14328: 34f — Wheeler Haines 412: 34c — Willkomm 433: 34e — Wimmer 155: 34f — Winkler 153: 34e — Wirtgen 706: 34a — de Wit 5215: 34e; 5237: 34a; 5273: 34a — Whitefoord 119: 23 — Woronow 273: 19.

Zahn 1471: 11 — Zohary & Amdursky 165: 27a.

II.9. INDEX OF TAXONOMIC NAMES

Accepted names in *Origanum* are in **bold type**; synonyms are in *italics*. To each new taxon, status, combination, and validated name a (!) is added. For the sub-chapters in chapter I only one reference for each taxon is made.